Theory of Games
and Strategies

RICHARD I. LEVIN
ROBERT B. DESJARDINS

Graduate School of Business Administration
University of North Carolina

INTERNATIONAL TEXTBOOK COMPANY

Scranton, Pa.

Copyright ©, 1970, by International Textbook Company

Library of Congress Catalog Card Number: 70—98510

Standard Book Number 7002   2252   9

*We dedicate this book to
two wonderful game players and strategists
Charlotte and Janet*

International's Series in Management Science

Theory of Games
and Strategies

Preface

This is a book about the general strategy of conflict—the examination of conflict situations and the mathematical development of strategies and responses in selected conflict situations. In it we will develop general rules concerning the logic that underlies strategic behavior of all types.

The book was not written for nor will it have any appeal to the professional mathematician; it demands only a modest background in mathematics and was not designed for the scientist whose interest is in pushing back the frontiers of knowledge. It was written instead for persons who want a clear and understandable yet sound and respectable approach to game theory beyond the introductory level.

The authors have attempted to avoid rigorous mathematical proofs and have structured the work so that the first four chapters can be read without a background in calculus. The material in the last three chapters requires a modest facility with differential calculus, which is presented as part of the exposition.

Chapter 1 is a general introductory unit in which we define terms, establish general conventions, and discuss the various methods of classifying games. The second chapter treats two-person zero-sum games where the number of choices open to each player is limited to two. In this unit the reader is introduced to the concepts of saddle points, pure strategies, mixed strategies, and game value. Chapters 3 and 4 are a continuation of Chapter 2 in which two-person zero-sum games of larger dimension are considered. These two units illustrate solution methods involving dominance, subgames, and for larger games, linear programming.

Chapter 5 introduces the reader to the concepts involved in non-zero-sum games—games in which the sum of the players' payoffs for some pairs of choices will not be zero. Issues involved in cooperation and bargaining are considered here also. Particular attention is given to the development of the ideas of status quo points, threat potentials, and security levels. The logic of Nash's solution is fully developed.

Chapter 6 treats games involving more than two players—n-person games. The equilibrium solution, and von Neumann's stable-set solution concept are developed. The final chapter introduces the reader to the concepts involved in nonnegotiated games. Here the psychological concept of rationality and its effect upon nonnegotiated games is treated in some detail.

A current bibliography of over a hundred and twenty-five items is included. Many of these entries are referenced in the chapters; others indicate the most logical sources of information for those wishing to pursue the area more thoroughly. In all, they represent a comprehensive bibliography for the subject area.

We have always considered game theory one of the more fascinating and agile areas of mathematics; it affords a rare opportunity to combine elements of behavioral and mathematical theory in a dynamic environment. We hope that this book will make the theory of games and strategies more understandable to those interested in it so that we may share with them the enjoyment we have had in writing it.

RICHARD I. LEVIN
ROBERT B. DESJARDINS

Chapel Hill, North Carolina
October, 1969

Contents

CHAPTER 1

Theory of Games

INTRODUCTION

In the sense that we shall use it in this book, the term "game" represents a conflict between two or more parties. Most of us are familiar with games like poker, bridge, and chess; however, the ideas we shall present in this book are more basic than the rules for playing any one of these common games. The ideas we shall explore are concerned with the *general* strategy of conflict—that is, the examination of conflict situations and the mathematical development of strategies and responses in selected conflict situations.

Game theory is really the "science of conflict," but mere knowledge of game theory will not make anyone a better poker player, bridge player, or more successful business manager. This is basically because game theory is not really concerned with finding an optimum or winning strategy for a particular conflict situation. Instead, it provides general rules concerning the logic that underlies strategic behavior of all types. Inasmuch as these rules are not too complex (at least not at the level on which we shall approach the subject here) one can easily be misled into believing that these rules indicate a sufficient set of decision rules for success in all situations of strategy. The danger in this belief is not that the rules we shall develop are fallacious, but rather that the situations in real life to which they are applicable are unfortunately few in number.

Game theory applies to a very limited number of situations which are technically known as "games." A more explicit definition of these situations would be cases where (a) there is a conflict of interests between the participants, (b) each of the players has several choices as to his appropriate actions, (c) the rules governing these choices are specified and known to all players, (d) the outcome of the game is affected by choices made by all of the players, and (e) the outcome for all specific sets of choices by all of the players is known in advance and numerically defined.

When a conflict situation meets these criteria, we can call it a game. On the other hand, there are many instances of conflict which are not games in the rigorous sense. An example of one of these would be a dispute between two persons. The motivation in this situation would be one of hostility, not logic. Generally these opponents do not intelligently calculate all of the alternative actions open to them, nor do they know in advance what particular outcome will always result from a stipulated set of actions and responses on their respective parts. We generally call these types of situations altercations or *fights* instead of games.

Game theory is concerned primarily with the logic of strategy. It is not concerned with the individual psychology of the players, nor with the respective intellectual strengths of the players. In this sense game theory can prescribe completely the strategy for tic-tac-toe quite easily, but falls short in defining a guaranteed winning approach to poker. In the former game all of the relevant moves (and counter-moves) are known to the players; in poker, chance and the ability of the players to create impressions concerning their strength (bluff) play a large part in the outcome of the game.

There are other reasons why game theory finds little application in real-life situations. First, game theory demands that numerical values be placed on the outcome of all possible sets of actions by the players of the game. For example, consider the case where two players match pennies. Let us suppose that one of these players wins when the coin faces are alike and loses when they are different. In this simple game, the numerical value to each of the players of each of the four possible outcomes (head-head, head-tail, tail-tail, and tail-head) can be calculated in advance and will not change during the play of this game. Consider now a more sophisticated game involving two competing companies, each with its own promotion campaign in a market. Who can really specify numerically the outcome when one of these "players" chooses a heavy commitment to television advertising at the same time that his competitor purchases significant space in local newspapers? In such a case the ability of game theory to provide real operational suggestions for appropriate strategies is at present very limited.

The second constraint to the applicability of game theory to business situations results from the theoretical concept of "winning." Game theory provides a set of strategies (or plays) which guarantee, in those situations where game theory is applicable, a certain outcome "in the long run." That is, if the game is played over an extended period of time in the same environment, we can specify what the long-run expectation of each of the players will be. However, in business (as well as in many other conflict situations) the environment changes in the long run, hence the concept of long-run gain has little meaning to the players. In this sense whenever the conflict situation will occur only once—i.e., whenever a single decision is called for from each of the players—game theory has minimal relevance.

The value of studying game theory is twofold. In one sense it can help us to recognize those situations where the rules of strategy *do* apply, thus allowing the participants to apply whatever rules of conflict *are* applicable as opposed to allowing the situation to remain at the level of an illogical fight. This substitution of intellectually determined strategies for emotionally based moves represents an achievement in itself. Another value of game theory is that it describes those situations where rules for strategic behavior *cannot* be applied, and thus by implication defines those areas which may be responsive to further inputs of scientific investigation. This scientific process of delineating the most fertile ground for investigation has certainly spawned its share of breakthroughs in the history of knowledge.

HISTORICAL DEVELOPMENT OF GAME THEORY

The theory of games of strategy was first proposed in 1921 by a French mathematician, Émile Borel. The first successful analysis and the accompanying proofs were offered by John von Neumann in 1928. Von Neumann's key contribution is known today as the proof of the minimax theorem, a concept we shall study in some detail in later chapters of this book. In 1944 the significant work in the field of game theory appeared, *The Theory of Games and Economic Behavior.* It was authored by von Neumann and a collaborating economist, Oskar Morganstern. The real significance of this book was that it represented one of those rare occasions in scientific publication where a new field was rather thoroughly explored in the first major work to be published in that field. In this sense von Neumann and Morganstern published their work about the same time that linear programming appeared on the scene. It was then recognized that game-theory problems could be formulated as special cases of linear programming; the elements of the simplex method of linear programming as proposed by George Dantzig were later used to prove the minimax theorem in game theory, and to provide solutions to games of large size. Since that time a significant library of books and articles on the subject of game theory has appeared in scientific literature. A representative sampling of these works appears in the Bibliography at the end of this book.

ELEMENTARY EXAMPLE OF A GAME

The basic concepts involved in game theory can be described by a simple game involving two persons; we shall call these two opponents X and Y. We can think of them as sitting across a table from each other, each with two buttons in front of him. We shall denote player X's buttons as a and b and player Y's buttons as r and s; thus each player has two choices open to him. We shall also suppose a partition between them such that neither can notice in advance which button his opponent is going to press. At a signal from a third party, each player presses one of his buttons. The results of each of the possible four combinations is known in advance to both of the players; the uncertainty inherent in the game arises from the fact that neither player knows what button his opponent will press next. Every time the third party signals, each player presses one of his buttons, and the game thus continues over time. At the end of, say, a hundred "plays," the game is over and the points won by each of the players are totaled and the winner is determined. It is assumed both players are of equal intelligence and that each actively attempts to win the game—that is, to amass the greatest number of points.

In Fig. 1-1, we have illustrated this particular simple game in

Fig. 1-1. Illustration of a simple game.

tabular form. The plays open to each of the opponents and the resulting winnings or losses (called *payoffs*) are indicated in the table. In this particular example we have been unfair to player Y—we have biased the game against him. One can reason to this conclusion by observing that if player X presses button a for each play of the game, player Y cannot win; in fact player Y faces a choice between losing 1 point on each play if he responds by pressing button r or losing two points each play if he responds by pressing button s. Of course, being intelligent, player Y will respond each time by pressing r, since this represents his least loss alternative; thus player X will win 1 point on each play of the game. We can say however, that even though player *Y* will lose this game, he has responded intelligently by minimizing his losses through choice of his most appropriate strategy (pressing button *r* each time).

The opposite situation—a game biased against player *X*—is illustrated in Fig. 1-2. In this case it is player *Y* who has been given the

Player *Y*

		Button *r*	Button *s*
	Button *a*	*X* wins 4 points	*Y* wins 2 points
Player *X*	Button *b*	*X* wins 3 points	*Y* wins 1 point

Fig. 1-2. Game biased against Player *X*.

advantage. *Y* observes that if he presses button *s* on each play of the game, *X* has no chance to win. Player *X* on the other hand, seeking to minimize his losses in this situation, counters by pressing button *b* on each play of the game, his better choice between two bad ones. Player *Y* wins a point each time the game is played in this simple illustration.

Not all games of this size have such simple solutions however. Figure 1-3 illustrates a game between two players where the determi-

Player *Y*

		Button *r*	Button *s*
	Button *a*	*X* wins 3 points	*Y* wins 4 points
Player *X*	Button *b*	*Y* wins 2 points	*X* wins 1 point

Fig. 1-3. A more complex game.

nation of individual strategies is more difficult. One can see from this game that player *X* would not press his button *a*, hoping to win 3 points each play simply because player *Y* could counter with button *s* and win 4 points himself. Neither would player *Y* press button *s* on every play hoping to win 4 points, since player *X* could counter with his button *b* and win 1 point himself. It is soon obvious from this game that it is advantageous for the players to play *each* of their choices (buttons) a part of the time. Just how to calculate these proportions of time to allot to each choice will be discussed in Chapter 2.

These illustrations are quite simple, and no involved thinking is required on the part of the players to determine the appropriate play and response. However, one can quickly imagine situations involving more than two players, each with many possible "buttons"; in these cases the number of possible choices becomes significantly large and the reasoning concerning proper plays becomes quite complex. These three initial illustrations of game situations are given only to introduce the reader to the basic idea of a "game"; later sections of this book treat in detail some of the more involved types of situations it is possible to encounter in game theory.

STANDARD CONVENTIONS FOR GAMES

In order to eliminate the necessity for written descriptions of the "payoffs" (winnings or losses for each possible combination of buttons), a standard set of conventions has been established in game theory. It is the usual practice to omit a description like "player *X* wins two points" and replace it with the integer 2. The positive algebraic sign which is assumed to accompany this number indicates that it is player *X* who benefits from this payoff. Similarly, instead of saying "player *Y* wins three points" one simply indicates this with the value – 3, the minus sign indicating that it is player *Y* who benefits from this particular payoff. As another standard convention we shall always let player *X* be the participant who has choices between the rows, and player *Y* will represent the player who chooses between the columns. These conventions are represented in Fig. 1-4.

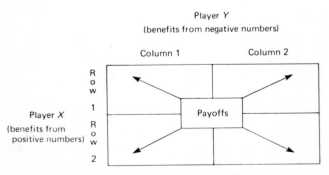

Fig. 1-4. Standard game conventions.

The use of the algebraic sign convention for payoffs in game theory is illustrated in Fig. 1-5. Here we have presented three differ-

	Player *Y*	
Player *X*	*X* wins 6 points	*Y* wins 2 points
	X wins 3 points	*Y* wins 1 point

	Y	
X	6	-2
	3	-1

	Player *Y*	
Player *X*	*X* wins 4 points	*Y* wins 1 point
	Y wins 3 points	*Y* wins 5 points

	Y	
X	4	−1
	−3	−5

	Player *Y*	
Player *X*	*Y* wins 7 points	*X* wins 2 points
	X wins 3 points	*Y* wins 5 points

	Y	
X	−7	2
	3	−5

Fig. 1-5. Algebraic sign conventions for games.

ent game situations, first using word descriptions of the payoffs; beside each of these we have illustrated the abbreviated algebraic sign conventions for the same games.

A further convention in game theory is known as *matrix notation*. In this case, games are represented in the form of a matrix (a rectangular array of numbers deriving from matrix algebra). When games are expressed in this fashion the resulting matrix is commonly called the "payoff matrix." Figure 1-6 gives several examples of games, illustrated first in rectangular table form; beside each of these games is its matrix equivalent. The third example in Fig. 1-6 illustrates a game in which each of the two opponents has three choices open to him. In this case we also notice that when player *Y* plays his third column and player *X* counters with his third row, the payoff for those particular choices is zero, indicating a draw.

The term *strategy* in game theory refers to the total pattern of choices employed by any player. For example, if player *X* chooses to play his first row half of the time and his second row half of the

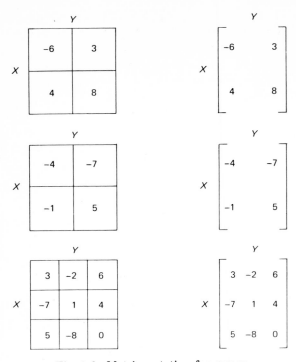

Fig. 1-6. Matrix notation for games.

time, his strategy for the game is $1/2, 1/2$. His strategy for each row can be referred to as being $1/2$.

GAMES DEFINED ACCORDING TO THE NUMBER OF PLAYERS

All of the examples up to this point have been of games involving two players. These types of games are commonly referred to as two-person games. In two-person games the players may have many possible choices open to them for each play of the game (the payoff matrix may have many columns and rows); however, the number of players is still two. Not all game situations are definable as two-person games; games can involve many people as active participants, each with his own set of choices for each play of the game. In Fig. 1-7 we have illustrated a pair of three-person games. In the first example, each of the three players has *two* choices open to him; in the second case, each player has three choices open to him. The complexity present in games involving three and more persons is quickly seen by observing that the addition of one more player increased the possible number of payoffs from 4 to 8 in the first exam-

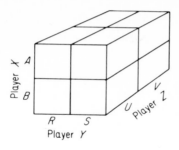

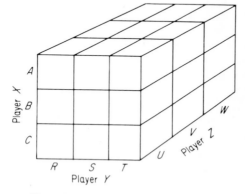

Fig. 1-7. Illustration of two three-person games.

ple, and from 9 to 27 in the second example. The mathematics involved in three-person and larger games is quite complex; Chapter 6 is devoted to an introduction to the ideas involved in games of large size with many players. It is interesting to note that the complexity of three-person and larger games derives not only from the absolute number of players but also from the fact that coalitions (temporary or permanent alliances between two or more players) can and do develop.

ZERO-SUM AND NONZERO-SUM GAMES

Games can be further classified concerning whether or not they are *zero-sum*. A zero-sum game is one in which the sum of the points won equals the sum of the points lost—that is, one player wins at the expense of the other(s). When two people match coins, for example, they are indulging in a zero-sum game, since whatever one of the players loses the other player wins. Poker, for another example, can be a zero-sum game or a nonzero-sum game. Often a percentage (or fixed amount) is withdrawn from each pot (the winnings from one play) for "the house," that is, to defray the overhead involved in promoting the game. In this case poker would be a nonzero-sum game. Where no deduction is made from the pot on each play, poker *is* an example of a zero-sum game.

There are however many examples of nonzero-sum games much more subtle than the preceding examples. A well-known example of a nonzero-sum game is the case of two competing firms each with a choice regarding its advertising campaign. If we assume that the market is quite small and geographically restricted it is quite reasonable to say that sales will remain at nearly the same level whether the firms engage in advertising or not. If neither firm advertises, they both save the cost of advertising and thus each receives a positive benefit, or payoff (representing a nonzero-sum outcome). In the event both firms spend equal amounts of money to advertise, neither firm's sales will increase (given equal quality of the advertising campaigns), and thus both firms will have incurred the cost of advertising without any benefits. Since both firms will lose money, this is clearly a nonzero-sum outcome. Even though this example has certain assumptions, and is therefore not really an operational illustration, it does serve to indicate that nonzero-sum games are frequently encountered in real business situations. Simply put, in zero-sum games the *sum* of the payoffs from any play of the game in the eyes of the players must be zero, whereas in nonzero-sum games it may be positive or negative. The analysis of nonzero-sum games is the major topic of Chapter 5.

VALUE OF A GAME

The term *value of a game* refers to the average payoff per play of the game over an extended period of time. In the case of the game illustrated in Fig. 1-8, for example, player *X* would play his first row

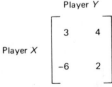

Fig. 1-8. A game
with a positive value.

on each play of the game and player *Y* would counter defensively by playing his first column each time, in order to minimize his losses.

Since player X wins three points on each play of the game, obviously his average winnings per play will also be 3 as long as the game is played. In this case, the fact that the value of the game is 3, with an implicit positive algebraic sign denotes that player X wins the game.

Another illustration is given in Fig. 1-9. Here player Y plays his

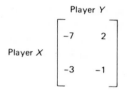

Fig. 1-9. A game with a negative value.

first column on each play of the game and player X counters defensively by playing his second row on each play of the game, thus minimizing his losses. We can see that Y wins 3 points on each play of the game. Since the average payoff per play is -3, the value of this game is -3, the minus sign indicating that Y is the winner. Determining the value of a game is of course not always this simple. In the case where the players determine that their best alternative is to play each row or each column a certain part of the time (such as in Fig. 1-3) calculation of the value of the game is a bit more complex. These situations are introduced in Chapter 2 of the book, along with several methods for calculating the value of the game.

OTHER MEANS OF CLASSIFYING GAMES

In addition to classifications concerning the number of players involved in the game, and classifications relative to whether the game is zero-sum or not, games may be further classified according to (a) the number of moves (or plays) involved in the game, and (b) the information content of the game. Although most of the illustrations involved in this book concern games with an unlimited number of moves, it is only fair to point out that in certain games the number of moves is limited to a fixed magnitude before play begins. In these kinds of games the players would certainly adopt strategies different from those they would employ if they knew that play could continue over an extended period of time. This is quite analogous to an investor with limited capital being asked on the one hand to "risk it all" on a single toss of a coin, as opposed to allowing him on the other hand to bet small portions of his capital on successive outcomes of a risky situation. His decision in either case would surely be different.

Games are further classified as to the information available to the players. In the case of chess, checkers, and games of this sort it is impossible to have secrets. Whatever strategy is adopted by either player can also be discovered by his competitor. These types of games are known as *games of perfect information.* On the other hand, there are games of imperfect information where neither player knows the entire situation and must be guided in part by his guess as to what the real situation is. Poker is certainly a game without perfect information, since the cards held by each player are not known until the end of the game, and sometimes not even then at the option of "cagy" players.

OUTLINE OF THE BOOK

Chapter 2 of the book covers simple two-person zero-sum games where there are only two choices open to each of the players. We demonstrate how to determine optimum strategies under these circumstances and how to calculate the value of the game. Chapter 3 treats two-person games where one of the players has more than two choices open to him. Two-person games in which there are three and more choices open to each of the players are the subject of Chapter 4; there we illustrate the use of linear programming in the solution of these larger games. Nonzero-sum games are covered in Chapter 5. Chapter 6 discusses n-person games, and the final chapter treats nonnegotiable games.

2 × 2 Two-Person Games

INTRODUCTION

In this chapter we introduce methods for determining optimum strategies for playing two-person zero-sum games with only two choices open to each player. These are denoted as 2 × 2 games. Once the fundamental ideas of strategy are developed, the chapters to follow apply these principles to games of larger size. The mathematics utilized in the determination of strategies will be limited to algebra. Once the strategies have been determined, we will demonstrate how the value of the game can be determined by either algebraic methods or by the use of simple probability ideas.

GAMES WITH A PURE STRATEGY

Figure 2-1 illustrates several games in which each of the players has two choices, i.e., player X can play either of the rows and player Y can play either of the columns. Beside each of the illustrations in Fig. 2-1 is a short description of how the players would determine their optimum strategies. Notice that in each of these cases, each of the players decides finally to limit his play to *one* of his rows or *one* of his columns. The players may experiment for a while but eventually both player X and player Y will select one alternative and will play it all the time assuming of course that each player desires to win, or if he cannot win, to minimize his losses.

When a player plays one row all of the time (or one column all of the time in the case of player Y), he is said to be playing a *pure strategy*. When one of the players elects to play a pure strategy, the other player will always logically counter with a pure strategy himself. . Figure 2-2 illustrates this point. Here we have illustrated a simple game in which player X would play his second row all of the

(a) X $\begin{bmatrix} 3 & 4 \\ 1 & -3 \end{bmatrix}$

Player X sees that if he plays his first row all of the time he cannot lose; thus he does so. Player Y counters by playing his first column all of the time, thereby limiting his losses to 3 points per play instead of 4.

(b) X $\begin{bmatrix} 7 & 6 \\ 1 & 3 \end{bmatrix}$

Player X sees that if he plays his first row all of the time, he wins more on each play, regardless of what player Y does (7 is preferable to 1 and 6 is preferable to 3). Player Y counters by playing his second column choosing to lose 6 points per play rather than 7.

(c) X $\begin{bmatrix} -8 & -7 \\ 4 & -2 \end{bmatrix}$

Player Y notices that if he plays his second column on each play, his opponent cannot win. Observing Y's choice, player X elects to play his second row all of the time, thereby limiting his losses to 2 points instead of 7.

(d) X $\begin{bmatrix} -1 & 5 \\ -3 & -2 \end{bmatrix}$

Player Y sees that to play his first column prevents his opponent from winning. Seeing that Y has made that choice, player X responds by playing his first row all of the time to limit his losses to 1 point per play.

Fig. 2-1. 2 X 2 games illustrated.

time, since to do so guarantees that his opponent cannot win. An intelligent opponent Y will obviously see that his best response in this case is to play his first column, thereby minimizing his losses (3 points per play loss instead of 5).

The same reasoning of course is used when the game is intentionally biased against player X. Figure 2-3 illustrates a game involving a pure strategy for player Y. In this case player Y will elect to play his first column on each play, since this strategy guarantees that he cannot lose. Player X must counter on each play by choosing to play his first row, thereby limiting his losses per play to 3 points

X $\begin{bmatrix} -1 & 2 \\ 3 & 5 \end{bmatrix}$

Fig. 2-2. Game with a pure strategy.

X $\begin{bmatrix} -3 & 4 \\ -6 & -2 \end{bmatrix}$

Fig. 2-3. Game with a pure strategy.

instead of 6. We see then that when one of the players elects to play a pure strategy, this automatically insures that his opponent will counter with a pure strategy (if his opponent wants to behave rationally), since one of the two choices open to his opponent will always be preferable to the other, unless of course they have identical values.

CONCEPT OF A SADDLE POINT

When a game involves a pure strategy on the part of the players, it is said to contain a *saddle point*. A saddle point is easily recognized because it is simultaneously the smallest algebraic value in its row *and* the largest algebraic value in its column.[1] The significance of this definition is apparent when we consider it in terms of the players' individual desires. Player *Y* would always rather have as his payoff the smallest algebraic value in any of the rows in the game. Figure 2-4 illustrates a 2 X 2 game with only one of the rows shown.

Fig. 2-4. Smallest algebraic value in the row.

Of the values in this row, player *Y* would choose as his better payoff the -6, since negative values represent payoffs to him, and -6 is the most negative value (the smallest algebraic value) in the row.

Looking at a game from *X*'s point of view we can reason why he would be attracted to the largest algebraic value in a column by remembering that positive values represent payoffs to him. As a

[1] The more proper terminology would be to call the payoff which is both a row minimum and a column maximum the *minimax* entry. The term *saddle point* when strictly used actually refers to the *row and column location of the minimax entry*. For example, in Fig. 2-3, -3 is the minimax entry; since this entry is found at the intersection of the first row and the first column, the row and column location indices (1-1) are actually the saddle point. It is common, however, to find the entry itself referred to as the saddle point, and as long as the reader understands this distinction, there is no harm in doing so.

simple illustration, Fig. 2-5 presents a 2 × 2 game with only one
column shown. In this case, player *X* would rather have as his payoff
the value 5, since it is the larger algebraic value in the column.

Fig. 2-5. Sad-
dle point illus-
trated.

Fig. 2-6. Sad-
dle point illus-
trated.

Naturally, when there is one payoff which is *both* the smallest
algebraic value in its row *and* the largest algebraic value in its column,
both players will choose that payoff and will play whatever row or
column is required to get it. Figure 2-6 illustrates a 2 × 2 game with
a saddle point. In this case, - 3 is *both* the smallest algebraic value in
the top row *and* the largest algebraic value in the second column (- 3
is of course larger than - 6 algebraically). The fact that it is the
smallest algebraic value in the row makes it more attractive to player
Y than 2 (the other alternative). Because it is the largest algebraic
value in its column it is more desirable to player *X* than - 6. Since it
satisfies both players, it will be the outcome of the game and thus the
value of the game—that is, player *Y* will win 3 points on each play of
this game.

Of course, not all two-person games have a saddle point; examination of the game matrix will quickly reveal whether or not one is
present. When a game has a saddle point, complex calculations to
determine the strategies for each player and the value of the game
are unnecessary. Figure 2-7 illustrates several games. In cases where
saddle points exist, they are circled and the strategies and value of
the game are noted. Strategies for those games where no saddle point
exists will be developed later in this chapter.

The choice of the term *saddle point* to represent pure strategy
situations was not accidental; in fact it is a highly descriptive term.
Anyone vaguely familiar with a horse saddle would agree with the
pictorial representation of two views of the contours of a saddle
shown in Fig. 2-8 (a view facing the horse and a view from the side of
a horse). In the view facing the horse we see a curve which is convex
in shape and from the side of the horse we notice a concave curve.
Directly under the rider there is a point which is the maximum point
on one curve (Fig. 2-8a) and at the same time the minimum point on

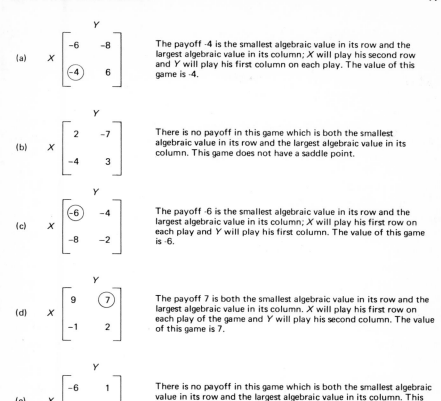

(a)   X
The payoff -4 is the smallest algebraic value in its row and the largest algebraic value in its column; *X* will play his second row and *Y* will play his first column on each play. The value of this game is -4.

(b)   X
There is no payoff in this game which is both the smallest algebraic value in its row and the largest algebraic value in its column. This game does not have a saddle point.

(c)   X
The payoff -6 is the smallest algebraic value in its row and the largest algebraic value in its column; *X* will play his first row on each play and *Y* will play his first column. The value of this game is -6.

(d)   X
The payoff 7 is both the smallest algebraic value in its row and the largest algebraic value in its column. *X* will play his first row on each play of the game and *Y* will play his second column. The value of this game is 7.

(e)   X
There is no payoff in this game which is both the smallest algebraic value in its row and the largest algebraic value in its column. This game does not have a saddle point.

Fig. 2-7. Determining a saddle point.

the other curve (Fig. 2-8b). In Fig. 2-9 this point is first shown individually on each of the two curves then on the combined curves.

The relationship of the saddle point to the strategies of the game players from this basis is quite simple. Player *X* is always desirous of having as a payoff the greatest algebraic value in a column of possible payoffs in any game. Player *Y*, on the other hand, always wants as

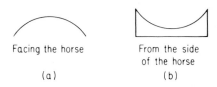

Facing the horse   From the side of the horse

(a)   (b)

Fig. 2-8. Two views of the contour of a saddle.

Fig. 2-9. Derivation of a saddle point.

his payoff the smallest algebraic value in a row representing possible payoffs in any game. When the payoffs are arranged in a game so that this point is common to two curves (Fig. 2-9c) it appeals to both players as their best choice, and thus they play to achieve this point. The term saddle point is also familiar to those with some experience in map reading. In this context it is defined as "a low point on terrain between two high points, which is also a high point between two low points." This situation can be visualized by thinking of the highest point of a valley located in a mountain range.

GAMES WITH MIXED STRATEGIES

In games where no saddle point exists, the players will resort to what is termed a mixed strategy, i.e., player X will play each of his rows a certain portion of the time and player Y will play each of his columns a certain part of the time. In business we have a close analogy. A manager follows a certain course of action A until an alternate course of action B appears to be more profitable. Later should A appear more attractive again, the manager switches back. The task for each player is to determine what proportion of the time he should spend on his respective rows and columns. In Fig. 2-10 we

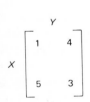

Fig. 2-10. Game with no saddle point.

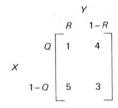

Fig. 2-11. Proportion designations.

have illustrated a simple two-person game with no saddle point; there are thus no pure strategies the players can use to play optimally. Since we are dealing here with the proportion of time player X plays each row and the proportion of time player Y plays each column, we shall indicate these proportions with letter symbols.

Suppose we let Q equal the fractional proportion (a number between zero and one) of the time player X spends playing the first row; then since 1 equals all of the time available for play, $1 - Q$ must equal the time he spends playing his second row. For example, if Q were found to be $^1/_3$, then player X would spend one-third of his time playing the first row and $1 - Q$ or $^2/_3$ of his time playing the second row. In the case of Player Y we shall let the proportion of the time he spends playing his first column be represented by the letter R. In this case $1 - R$ is the proportion of the time he spends playing his second column. Figure 2-11 illustrates the game introduced in Fig. 2-10 with the proper letter designations representing proportions of time spent on each column and row added.

An interpretation of Fig. 2-11 would indicate that

1. Player X plays the first row Q of the time (Q being a fraction between 0 and 1).
2. Player X plays the second row $1 - Q$ of the time ($1 - Q$ being a fraction between 0 and 1).
3. Player Y plays the first column R of the time (R being a fraction between 0 and 1).
4. Player Y plays the second column $1 - R$ of the time ($1 - R$ being a fraction between 0 and 1).

Our task is now to solve for the unknown fractions Q and R. Let us begin first by analyzing player X's situation. In short, player X wants to devise a strategy that will maximize his winnings (or minimize his losses) regardless of countermoves by his opponent Y. Logically X wants to divide his play between his rows so that his *expected* winnings[2] or losses when Y plays the first column will equal his expected winnings or losses when Y plays the second column. Of course, player Y (assumed to be as intelligent as player X), follows similar logic in calculating what proportion of time to spend on each of his columns. We must remember, however, that if player X was incompetent and followed a stupid strategy Y would simply look for the loopholes in X's strategy and play accordingly.

Figure 2-12 represents the expected winnings of player X from playing his first row Q of the time and his second row $1 - Q$ of the time. To make X's expected winnings when Y plays his first column equal to X's expected winnings when Y plays his second column, we

---

[2] *Expected* winnings is used here to indicate the sum, over time, of (the payoffs that will obtain) $\times$ (the probabilities that these payoffs will obtain). A simple example of the notion of "expectation" would be the case where there is one chance in 10 of hitting oil if we drill in a certain spot and where the value of a well if we hit oil is $1 million. The expected value of this drilling venture is then (.1 $\times$ $1 million) or $100,000.

X's Expected Winnings

	When Y plays column 1	When Y plays column 2
X plays row one Q of the time	X wins 1 point Q of the time	X wins 4 points Q of the time
	+	+
X plays row two 1 − Q of the time	X wins 5 points (1 − Q) of the time	X wins 3 points (1 − Q) of the time
	=	=
X's total expected winnings	1Q + 5 (1 − Q)	4Q + 3 (1 − Q)

Fig. 2-12. Expected winnings of player X.

equate these two expectations and solve for the appropriate Q value which does in fact make them equal. This calculation is performed as follows.

$$1Q + 5(1 - Q) = 4Q + 3(1 - Q)$$
$$Q + 5 - 5Q = 4Q + 3 - 3Q$$
$$5 - 4Q = Q + 3$$
$$- 5Q = - 2$$
$$Q = {}^2/_5$$

Therefore

$$(1 - Q) = {}^3/_5$$

Our algebraic solution indicates that to maximize his winnings from this game, player X should play his first row $^2/_5$ of the time and his second row $^3/_5$ of the time.

Now we must determine the appropriate strategies for player Y, using the same algebraic method. Player Y desires to divide his time between the columns so that no matter what X does about the rows, Y's expected winnings (or losses) when X plays row 1 will equal his expected winnings (or losses) when X plays row 2. In this manner player Y maximizes his winnings (or minimizes his losses) regardless of X's choice of strategy. These observations about Y's strategy are represented in algebraic form in Fig. 2-13. This figure indicates Y's expected payoff when he plays column 1, R of the time and column 2, $(1 - R)$ of the time, regardless of his opponent's actions.

Equating Y's expected losses when X plays row 1 with Y's expected losses when X plays row 2, and solving for the R value which

Y's Expected Losses

	Y plays column one R of the time		Y plays column two (1 – R) of the time	Y's total expected losses
When X plays row 1	Y loses 1 point R of the time	+	Y loses 4 points (1 – R) of the time	= 1R + 4(1 – R)
When X plays row 2	Y loses 5 points R of the time	+	Y loses 3 points (1 – R) of the time	= 5R + 3(1 – R)

Fig. 2-13. Expected losses of player Y.

produces an equality, we get

$$1R + 4(1 - R) = 5R + 3(1 - R)$$
$$R + 4 - 4R = 5R + 3 - 3R$$
$$4 - 3R = 2R + 3$$
$$- 5R = - 1$$
$$R + \,^1/_5$$

Therefore

$$(1 - R) = \,^4/_5$$

Our algebraic solution indicates to us that player Y, in order to minimize his expected losses in this case, should play column 1 one-fifth of the time and column 2 four-fifths of the time. Figure 2-14 shows the original game with the appropriate strategies for each of the players illustrated.

$$
\begin{array}{c}
Y \\
\begin{array}{cc}
\frac{1}{5} & \frac{4}{5}
\end{array} \\
X
\begin{array}{c}
\frac{2}{5} \\
\frac{3}{5}
\end{array}
\left[
\begin{array}{cc}
1 & 4 \\
5 & 3
\end{array}
\right]
\end{array}
$$

Fig. 2-14. Strategies for both players.

In the game we have used for illustration purposes, player X is the winner since the game is biased toward him (all of the payoffs are positive). In other cases, where there is an opportunity for player Y to win the game, both players would use the same method we have illustrated in determining their optimum strategies.   Figure 2-15

$$
\begin{array}{c}
 & Y \\
 & R \quad\quad 1-R \\
X \quad
\begin{array}{c} Q \\ 1-Q \end{array}
\begin{bmatrix} -3 & 2 \\ 4 & -5 \end{bmatrix}
\end{array}
$$

Player X's Expectations

$-3Q + 4(1-Q)$
$2Q - 5(1-Q)$

(expectations equated)

$-3Q + 4 \ (1-Q) = 2Q - 5(1-Q)$

(solution for proportions)

$$
\begin{aligned}
-3Q + 4 - 4Q &= 2Q - 5 + 5Q \\
-7Q + 4 &= 7Q - 5 \\
-14Q &= -9 \\
Q &= \frac{9}{14}
\end{aligned}
$$

therefore $1 - Q = \dfrac{5}{14}$

Player Y's Expectations

$-3R + 2(1-R)$
$4R - 5(1-R)$

(expectations equated)

$-3R + 2(1-R) = 4R - 5(1-R)$

(solution for proportions)

$$
\begin{aligned}
-3R + 2 - 2R &= 4R - 5 + 5R \\
-5R + 2 &= 9R - 5 \\
-14R &= -7 \\
R &= \frac{1}{2}
\end{aligned}
$$

therefore $1 - R = \dfrac{1}{2}$

Fig. 2-15. Strategies for both players.

illustrates such a game with the algebraic determination of both players' strategies shown in detail.

SIMPLIFIED METHOD FOR DETERMINING STRATEGIES

It is not necessary to solve two equations each time one desires to calculate the proper strategies for the players. Fortunately, much simpler solution methods have been devised which students of mathematics will recognize as being based on the use of determinants for the solution of equations. To illustrate just how they work, we shall take the game first introduced in Fig. 2-10, and apply this simpler technique to it. Figure 2-16 shows this game with the four solution steps illustrated. While this method does save considerable time, unfortunately it is limited to games involving a maximum of two choices for each player.

The same procedure also applies to games involving negative payoffs. For example, Fig. 2-17 illustrates just such a case.

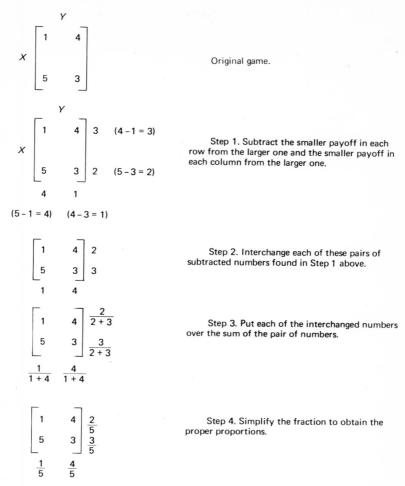

Original game.

Step 1. Subtract the smaller payoff in each row from the larger one and the smaller payoff in each column from the larger one.

Step 2. Interchange each of these pairs of subtracted numbers found in Step 1 above.

Step 3. Put each of the interchanged numbers over the sum of the pair of numbers.

Step 4. Simplify the fraction to obtain the proper proportions.

Fig. 2-16. Simplified solution illustrated.

PROPER METHOD OF PLAY

When the proportion of time each player should spend playing each of his rows or columns has been determined, the players must adopt this strategy in such a manner that neither can anticipate the order of play that will be followed by the other. The division of time between the rows or the columns by the respective players must be accomplished in a random fashion, that is, without any discernible pattern. If one of the players, for example, has found that the opti-

$$X \begin{bmatrix} -7 & 3 \\ 2 & -5 \end{bmatrix}$$

Y (above matrix)

Original game.

$$X \begin{bmatrix} -7 & 3 \\ 2 & -5 \end{bmatrix} \begin{matrix} 10 & ((3-(-7)=10)) \\ \\ 7 & ((2-(-5)=7)) \end{matrix}$$

Y (above matrix)

1. Payoffs subtracted.

9   8

((2-(-7) = 9))   ((3-(-5) = 8))

$$X \begin{bmatrix} -7 & 3 \\ 2 & -5 \end{bmatrix} \begin{matrix} 7 \\ \\ 10 \end{matrix}$$

Y (above matrix)

2. Pairs interchanged.

8   9

$$X \begin{bmatrix} -7 & 3 \\ 2 & -5 \end{bmatrix} \begin{matrix} \dfrac{7}{7+10} \\ \\ \dfrac{10}{7+10} \end{matrix}$$

Y (above matrix)

3. Pairs over the sum.

$\dfrac{8}{8+9}$ $\dfrac{9}{8+9}$

$$X \begin{bmatrix} -7 & 3 \\ 2 & -5 \end{bmatrix} \begin{matrix} \dfrac{7}{17} \\ \\ \dfrac{10}{17} \end{matrix}$$

Y (above matrix)

4. Fractions simplified.

$\dfrac{8}{17}$ $\dfrac{9}{17}$

**Fig. 2-17. Simplified solution method for games
with negative payoffs illustrated.**

mum division of time between his rows is to spend half of the plays
on each row, and he begins to play in the order: row 1, row 2, row 1,
row 2, etc., his opponent will simply adjust his strategy to take
advantage of the obvious advance disclosure of pattern, and probably
win considerably more than he would otherwise.

An example of this improper disclosure of pattern is shown in Fig. 2-18. Here we have illustrated a game with the optimum strate-

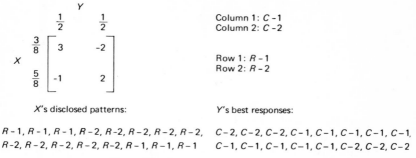

Column 1: C -1
Column 2: C -2

Row 1: R – 1
Row 2: R – 2

X's disclosed patterns:

R-1, R-1, R-1, R-2, R-2, R-2, R-2, R-2,
R-2, R-2, R-2, R-2, R-2, R-1, R-1, R-1

Y's best responses:

C-2, C-2, C-2, C-1, C-1, C-1, C-1, C-1,
C-1, C-1, C-1, C-1, C-1, C-2, C-2, C-2

Fig. 2-18. Disclosed patterns of play.

gies for each of the players determined. Using player X as an example, we have illustrated two disclosed patterns he might accidentally play. In both of these cases his opponent Y would depart from *his* optimally determined strategies and play the ones illustrated; that is, in neither case illustrated would Y play a $1/2$, $1/2$ strategy.

To avoid disclosing one's pattern of play, and to minimize the risk of winning less than one should (or losing more), a simple random choice method should be utilized. For example, using the optimum strategy determined for X in Fig. 2-18, ($3/8$ of the time spent playing the first row and $5/8$ of the time spent playing the second row), player X could put eight poker chips in a hat, 3 red and 5 white. His play on each turn would depend on which chip he drew, the red chip indicating the first row, the white chip the second row. Each time a chip was drawn, it would be replaced before another draw was made. Player Y would adopt the same random method, although he could "make do" with only two chips (one red and one white), since his optimum proportions are $1/2$, $1/2$.

DETERMINING THE VALUE OF THE GAME

Now that we have solved for the optimum strategies for both of the players, we can calculate the value of the game. The game first described in Fig. 2-10 together with the determined strategies is presented again in Fig. 2-19. If we look at the game first from player X's point of view, we reason as follows:

1. During the $1/5$ of the time that Y plays column 1, X wins 1 point $2/5$ of the time (when X plays row 1) and 5 points

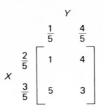

Fig. 2-19. Game
with strategies for
players.

$3/5$ of the time (when X plays row 2), and

2. During the $4/5$ of the time that Y plays column 2, X wins 4 points $2/5$ of the time (when X plays row 1) and 3 points $3/5$ of the time (when X plays row 2).

The total expected winnings of player X are then the algebraic sum of statements 1 and 2 above, or

$$1/5 \left[1(2/5) + 5(3/5) \right] + 4/5 \left[4(2/5) + 3(3/5) \right]$$

$$1/5 (2/5 + 15/5) + 4/5 (8/5 + 9/5)$$

$$1/5 (17/5) + 4/5 (17/5)$$

$$= 17/5 = \text{value of the game}$$

This means that if player X plays the strategy we have determined for him, he can expect to win an average payoff of $17/5$ points for each play of the game. Since the algebraic sign of the game value is positive, we know that X is the winner of this game; if on the other hand the value of the game determined above had been negative, this would signify that Y was the winner.

The game value could also be determined by looking at the situation from the point of view of player Y; he would reason as follows:

1. During the $2/5$ of the time that X plays row one, Y loses 1 point $1/5$ of the time (when Y plays column 1) and 4 points $4/5$ of the time, (when Y plays column 2), and

2. During the $3/5$ of the time that X plays row two, Y loses 5 points $1/5$ of the time (when Y plays column 1) and 3 points $4/5$ of the time, (when Y plays column 2).

The algebraic sum of statements 1 and 2 is

$$2/5 \left[1(1/5) + 4(4/5) \right] + 3/5 \left[5(1/5) + 3(4/5) \right]$$

$$2/5 (1/5 + 16/5) + 3/5 (5/5 + 12/5)$$

$$2/5 (17/5) + 3/5 (17/5)$$

$$= 17/5 = \text{value of the game}$$

Again we see that the value of the game is $1\frac{7}{8}$; because it is a positive number, we realize that X wins the game. The use of the term value of the game, does not mean that X wins $1\frac{7}{8}$ points on each play of the game; rather it signifies that X's average winnings per play over a considerable number of plays is $1\frac{7}{8}$ points.

In cases where games contain both negative and positive signs, the same procedure introduced above is followed to determine the game value. In the game in Fig. 2-20, there are payoffs with both

$$Y$$

$$X \quad \begin{array}{c} \frac{15}{23} \\ \\ \frac{8}{23} \end{array} \begin{bmatrix} \overset{\frac{11}{23}}{-5} & \overset{\frac{12}{23}}{3} \\ 7 & -8 \end{bmatrix}$$

Fig. 2-20. Game with
positive and negative
payoffs.

signs. The proper strategies have been calculated and are illustrated. If we want to evaluate the game from player Y's point of view, Y would reason as follows:

1. During the $\frac{15}{23}$ of the time X plays row one, Y will win 5 points $\frac{11}{23}$ of the time (when he plays column one) and lose 3 points $\frac{12}{23}$ of the time (when he plays column 2), and
2. During the $\frac{8}{23}$ of the time X plays his second row, Y will lose 7 points $\frac{11}{23}$ of the time, (when he plays column 1) and will win 8 points $\frac{12}{23}$ of the time (when he plays column 2).

The algebraic sum of these expectations is as follows:

$$\tfrac{15}{23}\left[-5\left(\tfrac{11}{23}\right)+3\left(\tfrac{12}{23}\right)\right]+\tfrac{8}{23}\left[7\left(\tfrac{11}{23}\right)-8\left(\tfrac{12}{23}\right)\right]$$

$$\tfrac{15}{23}\left(-\tfrac{55}{23}+\tfrac{36}{23}\right)+\tfrac{8}{23}\left(\tfrac{77}{23}-\tfrac{96}{23}\right)$$

$$\tfrac{15}{23}\left(-\tfrac{19}{23}\right)+\tfrac{8}{23}\left(-\tfrac{19}{23}\right)$$

$$=-1\tfrac{9}{23}=\text{value of the game}$$

It is possible to further simplify the procedure for finding the game value by remembering the logic behind the determination of the strategies themselves. In the case of player X, we determined a strategy that guaranteed X the same winnings regardless of his opponents' choice of columns, i.e., X won the same number of points (over time) when Y played his first column as he did when Y shifted to his second column. Thus it is possible to determine the value

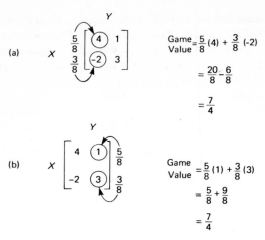

Fig. 2-21. Simplified method for calculating game value.

of the game by simple multiplication, as demonstrated in Fig. 2-21. Since player X expects to win as much when Y plays column 1 as he does when Y plays column 2, whatever X wins from either column will be identical and will equal the value of the game; this is supported in Fig. 2-21; the game value can be calculated by performing *either* of the paired multiplications.

The same reasoning can be applied to Y's expectations. He plays a strategy designed such that he wins the same points when X plays row 1 as he does when X plays row 2. The game in Fig. 2-22 is the

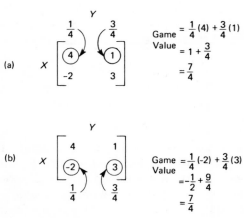

Fig. 2-22. Simplified method for calculating game value.

same as that introduced in Fig. 2-21; in this case, however, we calculate the value of the game from player Y's point of view, considering either row, that is, using either of the two paired multiplications we wish; both yield the same answer for the value of the game.

In the game presented in Fig. 2-23, we obtain the same answer

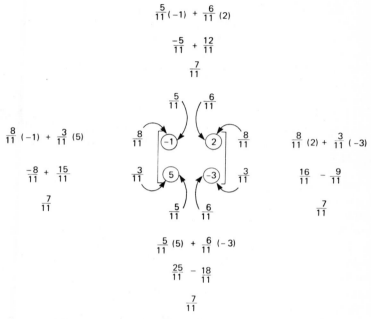

Fig. 2-23. Value of the game—four approaches.

for the game value using four different approaches; we include X's expectations when Y plays column one; X's expectations when Y plays column two; Y's expectations when X plays row 1, and finally Y's expectations when X plays row 2; in all four cases the answer is the same. Any *one* of these calculations is sufficient to determine the value of the game.

In each 2×2 game without a saddle point, each player's strategy consists of two probabilities denoting that portion of the time he spends on each of his rows or columns. Since each player plays a random pattern, and since neither knows what the other will play for the next move, the choice of rows for player X is statistically independent of the choice of columns for player Y. In the game illustrated in Fig. 2-24, the payoffs $(5, -7, -6, -4)$ are attained only if a particular column and a particular row are played simultaneously.

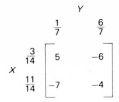

Fig. 2-24. Game il-
lustrating probabili-
ties of payoffs.

For example, the probability that *row 1* and *column 2* will be played
simultaneously is a joint probability. Under conditions of statistical
independence, $P(AB) = P(A) \times P(B)$, and is calculated; $3/_{14} \times {}^6/_7$. The
probability that – 6 will be the payoff on any one play is then $^9/_{49}$.

Using this reasoning, we can calculate the joint probabilities that
each of the four possible payoffs will be obtained. This has been
done in Fig. 2-25.

Payoff	Strategies Which Produce This Payoff	Probability of This Payoff
5	Row 1, column 1	3/14 X 1/7 = 3/98
-7	Row 2, column 1	11/14 X 1/7 = 11/98
-6	Row 1, column 2	3/14 X 6/7 = 18/98
-4	Row 2, column 2	11/14 X 6/7 = 66/98
		sum = 1.0

Fig. 2-25. Probabilities of each payoff.

In Fig. 2-26 we have computed the value of this game by multiplying each of the payoffs by the probability that it will occur in
random play, and summing the results.

Payoff		Probability That This Payoff Will Occur		
5	X	3/98	=	15/98
-7	X	11/98	=	-77/98
-6	X	18/98	=	-108/98
-4	X	66/98	=	-264/98
		total	=	$-4\frac{3}{7}$ = (Game value)

Fig. 2-26. Determining the value of the game.

SUMMARY OF 2 X 2 GAMES

Beginning with simple two-person games which contain a saddle
point, we have seen that these games are strictly determined—their

outcomes are determined in advance and neither player can do better than the value of the game (assuming that both play intelligently). In games without saddle points it is still possible to determine the best mixed strategy for each player; once this has been done, again the outcome is determined. Neither player is able to do any better than the game value under the same assumptions concerning the intelligence of the players. In this second case, however, determination of optimal mixed strategies assumes also that the payoffs in the game matrix have been assigned on an interval scale—that is, that a payoff of 4 to player X represents a condition which in his own individual preferences is exactly twice as good as a payoff of 2. If either of the players does not have this interval utility preference, (if for example player Y thinks a payoff of -3 is half as good as a payoff of -8), then it is impossible to determine mixed strategies in the manner described.

CHAPTER 3

2 × m and m × 2 Games

INTRODUCTION

A game in which one of the players has more than two choices of rows or columns and in which the other player has exactly two choices is referred to as an $m \times 2$ or $2 \times m$ game, depending on whether it is the player who plays the rows or the columns who has more than two choices. Figure 3-1 illustrates several of these games

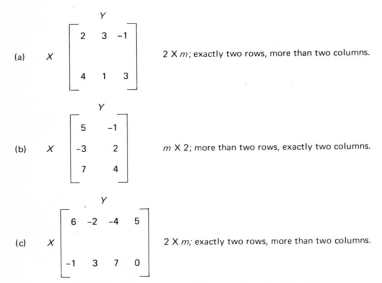

(a) X $\begin{bmatrix} 2 & 3 & -1 \\ \\ 4 & 1 & 3 \end{bmatrix}$ 2 X m; exactly two rows, more than two columns.

(b) X $\begin{bmatrix} 5 & -1 \\ -3 & 2 \\ 7 & 4 \end{bmatrix}$ m X 2; more than two rows, exactly two columns.

(c) X $\begin{bmatrix} 6 & -2 & -4 & 5 \\ \\ -1 & 3 & 7 & 0 \end{bmatrix}$ 2 X m; exactly two rows, more than two columns.

Fig. 3-1. Examples of 2 X m and m X 2 games.

with the proper dimensions indicated beside each of them. In these type games, one of the players may have many possible rows or columns as long as the other player is limited to two choices. In the formal literature of game theory, it is not the normal practice to break down the classification of games into 2 × *m* or *m* × 2; usually these games are considered to be just another type of matrix game. There is considerable practical value, however, in making this classification, since there are particular solution methods for games which are illustrated best by using 2 × *m* and *m* × 2 games as examples. For this reason we have elected to use classification by size as a pedagodical device in this book.

SOLUTION BY SADDLE POINT

2 × *m* and *m* × 2 games may have saddle points, and when this is the case, the proper strategies for each of the players can quickly be determined. Each of the three games illustrated in Fig. 3-2 contains a saddle point. It has been circled and the strategies for both of the players have been given together with the value of the game. Each of the games in Fig. 3-2 has a pure strategy which is best for both of the players.

(a)   X

$$\begin{bmatrix} 4 & 9 & \boxed{-3} \\ & & \\ 2 & -6 & -7 \end{bmatrix}$$

The value-3 is both the smallest value in its row and the largest value in its column; player *X* would play his first row all of the time and player *Y* would counter by playing his third column all of the time. The value of this game is-3.

(b)   X

$$\begin{bmatrix} \boxed{6} & 8 \\ -5 & 2 \\ 3 & -4 \end{bmatrix}$$

The value 6 is a row minimum and a column maximum; player *X* would play row one all of the time and player *Y* would play column one all of the time. The value of this game would be 6.

(c)   X

$$\begin{bmatrix} -4 & -2 & 1 & 5 & 9 \\ & & & & \\ \boxed{-2} & -1 & 0 & 4 & 8 \end{bmatrix}$$

The value-2 is a row minimum and a column maximum; player *X* would play his second row on all of the plays and player *Y* would counter by playing his first column. The value of this game is-2.

Fig. 3-2. Illustrations of saddle points.

SOLUTION BY DOMINANCE

When a 2 × *m* or *m* × 2 game does not contain a saddle point, it is still possible to solve for the strategies by using a method known as

Fig. 3-3. Illus-
tration of domi-
nance.

dominance. The game in Fig. 3-3 will introduce this concept. Since column one contains two positive payoffs, what advantage is there for player Y to consider playing this column? Column 2 with its two negative payoffs would certainly be more attractive to Y using this reasoning: When X plays his first row Y prefers - 1 to 4 as a payoff, and when X plays his second row Y certainly would rather have - 2 as a payoff instead of 7. Thus *regardless* of the row X chooses to play, Y always prefers his second column to his first. We can then say that the first column is *dominated* by the second column, and would never be played by Y. As soon as Y decides never to play his first column, the game is reduced to size 2 X 2; the new game appears in Fig. 3-4. The solution to this reduced game can be found by using the methods introduced in Chapter 2. We first determine if the reduced game contains a saddle point, and finding that there is none, we solve the game using the algebraic method. The strategies and the value of this reduced game appear also in Fig. 3-4.

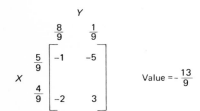

Fig. 3-4. Reduced game from
Fig. 3-3.

$$\text{Value} = -\frac{13}{9}$$

Fig. 3-5. Sec-
ond example
of dominance.

When a particular row or column is discarded because of dominance, the player making that decision discards it, because some other row or column will *always* return that player a better payoff *regardless* of his opponent's actions. In Fig. 3-5 we have illustrated a second example of dominance. We first determine whether there is a saddle point in this game; finding none, we proceed to see if there is a clear case of dominance.

We shall begin the analysis by comparing row 2 with row 1. When *Y* plays his first column, *X* prefers a payoff of 7 (row 1) to a payoff of - 1, (row 2); but when *Y* is playing his second column, *X* would prefer a payoff of 6 (row 2) to one of 4 (row 1). Thus as *Y* shifts his play from column to column, neither row 1 nor row 2 is *always* preferred by player *X*. There is no clear case of dominance between *these* two rows.

Let us now compare the second row with the third. When *Y* plays the first column, *X* would rather have a payoff of 5 (row 3) than one of - 1 (row 2); and when *Y* plays the second column, *X* prefers a payoff of 9 (row 3) in preference to one of 6 (row 2). Thus regardless of what *Y* does, *X* prefers his third row to his second, and discards the second row as an alternative. Row 2 has been dominated by row 3, and the game has now been reduced to 2 × 2 The reduced game appears in Fig. 3-6. Since there is no saddle point in the re-

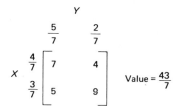

Fig. 3-6. Reduced game from Fig. 3-5.

duced game an algebraic solution is required. The strategies and the value of the game appear also in Fig. 3-6.

Figure 3-7 illustrates several additional examples of dominance. In each case we indicate the original game, then give the reasoning that would allow the game to be reduced by dominance. The resulting 2 × 2 game with the strategies and the game value is then presented.

SOLUTION BY METHOD OF SUBGAMES

Many 2 × *m* or *m* × 2 games can be reduced by dominance to a 2 × 2 game which can then be solved using the algebraic method for determining the strategies. There are however many other cases where such reduction is impossible or only partially successful. When such is the case, the reduced game is still larger than 2 × 2; in these

$$
\text{(a)} \quad X \quad
\begin{array}{c}
Y \\
\left[
\begin{array}{ccccc}
6 & 3 & -1 & 0 & -3 \\
3 & 2 & -4 & 2 & -1
\end{array}
\right]
\end{array}
$$

Player Y chooses not to play columns 1, 2, or 4, since either column 3 or column 5 (both with two negative payoffs) offers a better alternative, regardless of X's actions. From Y's point of view a column with two negative payoffs would always dominate any column which contained two positive payoffs.

$$
X \quad
\begin{array}{cc}
 & Y \\
\begin{array}{c} \\ \frac{3}{5} \\ \frac{2}{5} \end{array} &
\begin{array}{cc}
\frac{2}{5} & \frac{3}{5} \\
\left[
\begin{array}{cc}
-1 & -3 \\
-4 & -1
\end{array}
\right]
\end{array}
\end{array}
$$

$$\text{Value} = -\frac{11}{5}$$

$$
\text{(b)} \quad X \quad
\begin{array}{c}
Y \\
\left[
\begin{array}{cc}
-2 & 4 \\
-1 & 4 \\
3 & 1
\end{array}
\right]
\end{array}
$$

X compares row 1 with row 2. He reasons that when his opponent plays column one, X prefers a payoff of -1, (a loss of 1) to a payoff of -2, (a loss of 2); when Y plays the second column, X is indifferent since both payoffs are identical. Because row 2 is preferred when Y plays column 1 and because X is indifferent between rows 1 and 2, when Y plays column 2, row 1 is dominated.

$$
X \quad
\begin{array}{cc}
 & Y \\
\begin{array}{c} \\ \frac{2}{7} \\ \frac{5}{7} \end{array} &
\begin{array}{cc}
\frac{3}{7} & \frac{4}{7} \\
\left[
\begin{array}{cc}
-1 & 4 \\
3 & 1
\end{array}
\right]
\end{array}
\end{array}
$$

$$\text{Value} = \frac{13}{7}$$

$$
\text{(c)} \quad X \quad
\begin{array}{c}
Y \\
\left[
\begin{array}{ccc}
3 & -5 & -5 \\
0 & 2 & 4
\end{array}
\right]
\end{array}
$$

When player Y compares column 2 with column 3, he finds himself indifferent as long as X is playing row 1, i.e., two payoffs each -5. However, when X is playing row 2, player Y prefers the payoff of 2 (a loss of 2 points) to the payoff of 4 (a loss of 4 points). Y reasons that if column 2 is better for him part of the time (when X plays row 2) and as good as column 3 the rest of the time, column 3 is dominated by column 2.

$$
X \quad
\begin{array}{cc}
 & Y \\
\begin{array}{c} \\ \frac{2}{10} \\ \frac{8}{10} \end{array} &
\begin{array}{cc}
\frac{7}{10} & \frac{3}{10} \\
\left[
\begin{array}{cc}
3 & -5 \\
0 & 2
\end{array}
\right]
\end{array}
\end{array}
$$

$$\text{Value} = \frac{3}{5}$$

Fig. 3-7. Examples of dominance.

cases the use of another method for solution is required. The game in Fig. 3-8 is an example of this situation. The original game of size

Player *Y* could reason that his second column dominates the first, third, and fifth columns; when *X* is playing his first row, the payoff 4 is more desirable to *Y* than either 9, 7, or 8. When *X* plays his second row, the payoff of 3 to *Y* is more desirable than either 6, 5, or 8. Thus *Y* will never play columns 1, 3 or 5.

Fig. 3-8. Reduction by dominance to 2 X 3.

2 X 6 can be reduced by dominance to one of size 2 X 3; the reasoning and the reduced game are both shown in the figure.

It is impossible to reduce the smaller game appearing in Fig. 3-8 any further by dominance, because there is no column which is *always* preferred by *Y* regardless of what his opponent might do. As long as *X* played row 1, for example *Y* would prefer column 1 to column 2, but when *X* shifted play to row 2, *Y* would prefer column 2 to column 1. If player *Y* compares column 3 with column 2, he finds a similar situation—that is, when *X* plays row 1, *Y* prefers column 3 but when *X* plays the second row, *Y* prefers column 2. And finally, when *Y* compares column 1 with column 3, the reasoning is the same. *Y* prefers column 3 when *X* plays row 1 but shifts his preference to column 1 when *X* shifts to the second row. Thus we generate from the original 2 X 6 game, a 2 X 3 game with no saddle point and no further reduction by dominance possible.

Player *Y*, since he can choose not to play one of his columns if he so desires, can actually think of the reduced 2 X 3 game as being three separate subgames, each of size 2 X 2. The original 2 X 3 game and the three 2 X 2 games appear in Fig. 3-9. If *Y* chooses not to play the first column for example, subgame (c) results; if he chooses not to play the third column, subgame (a) results, and of course if *Y* elects not to play his second column, subgame (b) results. Thus *Y* can unilaterally determine just which 2 X 2 subgame *will* be played. Each of the subgames has a game value, and, of course, *Y* being the player with the choice in this example will pick that particular sub-

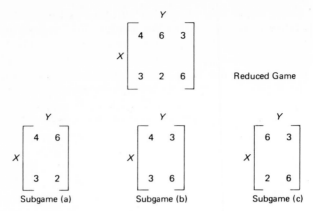

Fig. 3-9. Subgames illustrated.

game with the lowest value—the one which represents the lowest possible loss to him. In Fig. 3-10 we have illustrated the three subgames and have computed the value of each of them. Subgame (a) contains a saddle point (4) which is the value of that game. The

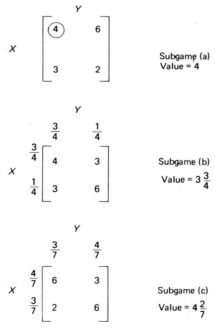

Fig. 3-10. Strategies and values
for subgames illustrated.

strategies of the other two subgames can be determined by using the algebraic method. Subgame (b) with a value of $3^3/_4$ points represents the lowest loss to player Y and should be the game he elects to play.

One might well question why playing some mixed strategy among the three columns of the 2 X 3 game would not represent a better alternative to Y than limiting his play to subgame (b). If Y did decide to play some mixed strategy using all three columns, he would really be playing subgame (a) part of the time, subgame (b) part of the time and subgame (c) part of the time. For example, if he decided to play each of his columns an equal part of the time, ($^1/_3$, $^1/_3$, $^1/_3$), he would in fact be playing each of the subgames an equal part of the time. No matter what proportions he decided upon, we could quickly calculate which part of the time each of the subgames was being played. Looking again at Fig. 3-10, we see that if player Y elects to play a mixed strategy using all three columns, he will expect to lose 4 points a part of the time, $3^3/_4$ points a part of the time and $4^2/_7$ points a part of the time. No matter what proportions he chooses for his mixed three-column strategy, he cannot reduce his losses to $3^3/_4$ points, which is what he expects to lose when he limits his play to subgame (b). For this reason, one of the subgames which result from a 2 X m or m X 2 game is *always* preferred by the player who makes the choice when compared to an all-column or all-row strategy.

In Fig. 3-11 we have illustrated another example of subgames.

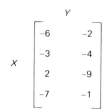

Fig. 3-11. 4 X 2
game illustrated.

This time, we have begun with a game of size 4 X 2 which cannot be further reduced by dominance. In this case six separate 2 X 2 subgames can be derived from the original game, either by breaking the game down into 3 X 2 games and these into 2 X 2 games or breaking the game directly into 2 X 2 games. Each of these subgames with the appropriate strategies and game value is illustrated in Fig. 3-12.

Since the value of all six subgames is negative, player X, who is making the choice, will naturally pick the game with the smallest negative value—subgame (e). In this case, X will play a two-row mixed

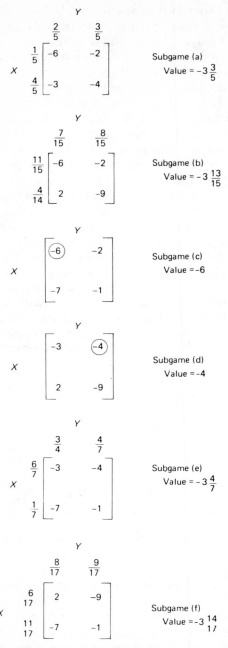

Fig. 3-12. Strategies and value
for six subgames.

strategy between the second and fourth rows of the original game and he will expect to lose an average of $3^4/_7$ points per play of the game, his minimum possible loss. For the same reasons described earlier, this choice of a mixed strategy between two rows represents a better alternative for X than any possible three- or four-row mixed strategy. A logical conclusion to be drawn from these examples is that two intelligent opponents will each play a strategy consisting of the same number of rows or columns. Given that one player consistently plays a two-row strategy, the best possible alternative for his opponent is to play a two-column strategy, in turn.

PROOF OF THE OPTIMUM STRATEGIES

It is not difficult to prove the logic of choosing one subgame as the best alternative. To help us here we have constructed a 3×2 game which appears in Fig. 3-13. In this instance player X has a

$$
\begin{array}{c}
Y \\
X \begin{bmatrix} 2 & 6 \\ 4 & 3 \\ 5 & 1 \end{bmatrix}
\end{array}
$$

Fig. 3-13. 3×2
game illustrated.

choice of which two rows to play; he makes this choice by evaluating the values of the three possible subgames, and then chooses the one with the largest positive value. In Fig. 3-14, we have illustrated each of the three subgames, including the strategies and the game value. Player X would choose to play subgame (a) which offers him a value of $3^3/_5$ points.

In Fig. 3-15 we have repeated the original game from Fig. 3-13 including the strategies for the optimum subgame (a). The strategy for the row which player X chooses not to play (row 3 of the original game) is indicated by a zero.

In Fig. 3-16 we have indicated the proportion of the time each player plays each of his rows or columns by the letters X and Y with appropriate subscripts. From our earlier explanation that a mixed strategy is chosen by equating expectations, we recall that X determined his strategies in such a way that he won (or lost) the same from playing each row regardless of Y's choice of column. Put another way, X expects to win the same number of points when Y plays

$$
\begin{array}{cc}
 & Y \\
 & \begin{array}{cc} \frac{3}{5} & \frac{2}{5} \end{array}
\end{array}
$$

(a) X $\begin{array}{c} \frac{1}{5} \\ \frac{4}{5} \end{array}$ $\begin{bmatrix} 2 & 6 \\ 4 & 3 \end{bmatrix}$ Value = $3\frac{3}{5}$

$$
\begin{array}{cc}
 & Y \\
 & \begin{array}{cc} \frac{5}{8} & \frac{3}{8} \end{array}
\end{array}
$$

(b) X $\begin{array}{c} \frac{1}{2} \\ \frac{1}{2} \end{array}$ $\begin{bmatrix} 2 & 6 \\ 5 & 1 \end{bmatrix}$ Value = $3\frac{1}{2}$

$$
\begin{array}{c}
 Y
\end{array}
$$

(c) X $\begin{bmatrix} 4 & ③ \\ 5 & 1 \end{bmatrix}$ Value = 3 (the saddle point)

Fig. 3-14. Subgames with strategies and values.

column 1 as he does when Y plays column 2. We will also remember that if he has chosen correctly, player X is guaranteed to win the value of the game, and *may win more*, in those cases where Y fails to play *his* optimum strategies.

These expectations of player X can be reduced to two mathematical statements using the symbolic notation introduced in Fig. 3-16. These are

X's Expectations

When Y plays column 1 $2X_1 + 4X_2 + 5X_3 \geqslant V$

When Y plays column 2 $6X_1 + 3X_2 + 1X_3 \geqslant V$

These statements are interpreted as follows: When Y plays his first column, X expects to win at least V, the value of the game, as long as

$$
\begin{array}{cc}
 & Y \\
 & \begin{array}{cc} \frac{3}{5} & \frac{2}{5} \end{array}
\end{array}
$$

X $\begin{array}{c} \frac{1}{5} \\ \frac{4}{5} \\ 0 \end{array}$ $\begin{bmatrix} 2 & 6 \\ 4 & 3 \\ 5 & 1 \end{bmatrix}$

Fig. 3-15. Strategies for the subgame which is played.

$$
\begin{array}{cc}
 & Y \\
 & \begin{array}{cc} Y_1 & Y_2 \end{array}
\end{array}
$$

$\begin{array}{c} X_1 \\ X_2 \\ X_3 \end{array}$ $\begin{bmatrix} 2 & 6 \\ 4 & 3 \\ 5 & 1 \end{bmatrix}$

Fig. 3-16. Symbolic designation of strategies.

he uses properly determined strategies, X_1, X_2, and X_3. He may of course win more if Y chooses poor strategies, hence the inequality symbol. Now let us replace the X subscript symbols with the strategies we found in Fig. 3-5 by using the method of subgames. Remembering that X's third row strategy is zero, we get

$$2(^1/_5) + 4(^4/_5) + 5(0) \geqslant 3^3/_5$$
$$6(^1/_5) + 3(^4/_5) + 1(0) \geqslant 3^3/_5$$

which reduces to

$$^2/_5 + {}^{16}/_5 \geqslant 3^3/_5$$
$$^6/_5 + {}^{12}/_5 \geqslant 3^3/_5$$

and finally to

$$3^3/_5 \geqslant 3^3/_5 \text{ (inequality is satisfied)}$$
$$3^3/_5 \geqslant 3^3/_5 \text{ (inequality is satisfied)}$$

Thus we observe that the strategies we have chosen for player X *do* support the guarantee implicit in the choice of mixed strategies.

Suppose we now test to see if the strategies we have determined for player Y support the game guarantee. Y's guaranteed expectations are

	Y's Expectations
When X plays row 1	$2Y_1 + 6Y_2 \leqslant V$
When X plays row 2	$4Y_1 + 3Y_2 \leqslant V$
When X plays row 3	$5Y_1 + 1Y_2 \leqslant V$

Here the inequality sign is reversed. This means simply that if we have chosen Y's strategies correctly, he will lose the value of game *but* may lose less if X plays poor strategies. Inserting Y's correct strategies from Fig. 3-15, we have

$$2(^3/_5) + 6(^2/_5) \leqslant 3^3/_5$$
$$4(^3/_5) + 3(^2/_5) \leqslant 3^3/_5$$
$$5(^3/_5) + 1(^2/_5) \leqslant 3^3/_5$$

which reduces to

$$^6/_5 + {}^{12}/_5 \leqslant 3^3/_5$$
$$^{12}/_5 + {}^6/_5 \leqslant 3^3/_5$$
$$^5/_5 + {}^2/_5 \leqslant 3^3/_5$$

and finally to

$$3^3/_5 \leqslant 3^3/_5 \text{ (inequality is satisfied)}$$
$$3^3/_5 \leqslant 3^3/_5 \text{ (inequality is satisfied)}$$
$$③^2/_5 \leqslant 3^3/_5 \text{ (inequality is satisfied)}$$

Once again all of the inequalities are satisfied by Y's strategies. We may infer then that we have correctly determined the optimum strategies for both players. Y's third inequality has a special message. We notice here that if X played his third row, Y would expect to lose only $3^2/_5$ points, instead of the game value ($3^3/_5$ points). Obviously the third row would be a poor choice for player X since it nets him $^1/_5$ point less than the game value. Our choice of subgame (a) (Fig. 3-14), in which the third row was *not* played by X, is completely supported by the inequality test.

The strategies for the *correct* subgame satisfied all five game inequalities. It is impossible for the strategies of the other two subgames to satisfy the game inequalities. We will demonstrate this with one example. In this case we have picked subgame (b) from Fig. 3-14. Player X's expectations again are

$$2X_1 + 4X_2 + 5X_3 \geqslant V$$
$$6X_1 + 3X_2 + 1X_3 \geqslant V$$

Substituting the strategies and value for subgame (b), we have

$$2(^1/_2) + 4(0) + 5(^1/_2) \geqslant 3^1/_2$$
$$6(^1/_2) + 3(0) + 1(^1/_2) \geqslant 3^1/_2$$

which reduces to

$$^2/_2 + ^5/_2 \geqslant 3^1/_2$$
$$^6/_2 + ^1/_2 \geqslant 3^1/_2$$

and finally to

$$3^1/_2 \geqslant 3^1/_2 \text{ (inequality is satisfied)}$$
$$3^1/_2 \geqslant 3^1/_2 \text{ (inequality is satisfied)}$$

The expectations of player Y are

$$2Y_1 + 6Y_2 \leqslant V$$
$$4Y_1 + 3Y_2 \leqslant V$$
$$5Y_1 + 1Y_2 \leqslant V$$

When we substitute the strategies determined for subgame (b) (Fig. 3-15) into these inequalities, we get

$$2(^5/_8) + 6(^3/_8) \leqslant 3^1/_2$$
$$4(^5/_8) + 3(^3/_8) \leqslant 3^1/_2$$
$$5(^5/_8) + 1(^3/_8) \leqslant 3^1/_2$$

which reduces to

$$^{10}/_8 + {}^{18}/_8 \leqslant 3^1/_2$$
$$^{20}/_8 + {}^9/_8 \leqslant 3^1/_2$$
$$^{25}/_8 + {}^3/_8 \leqslant 3^1/_2$$

and finally to

$$3^1/_2 \leqslant 3^1/_2 \text{ (inequality is satisfied)}$$
$$3^5/_8 \leqslant 3^1/_2 \text{ (this inequality is violated;}$$
$$\boxed{3^5/_8} \text{ is greater than } 3^1/_2)$$
$$3^1/_2 \leqslant 3^1/_2 \text{ (inequality is satisfied)}$$

The strategies from the incorrect subgame are thus unable to satisfy the inequalities which represent the game guarantee to the players. In particular, the second inequality directly above (representing Y's expectation when X plays the second row of the original game) cannot be satisfied by subgame (b). If X plays subgame (b) (containing the row which is omitted in the optimum subgame for him), then Y expects to lose more than the game value, a condition that violates the guarantee to Y implicit in the method we have used to determine mixed strategies. If we tested the strategies for subgame (c), Fig. 3-15, in the game inequalities, we would find also that one or more of the inequalities was not satisfied.

When the mixed strategies for each of the players are determined using the methods we have outlined thus far, each player is guaranteed at least a minimum position. X must win *at least* the value of the game, and Y cannot lose more than the value of the game. Unless both of these guarantees are satisfied by the strategies that result, we can infer that we have not determined the best strategies available to the players.

CHAPTER 4

3 × 3 and Larger Games

INTRODUCTION

Many of the solution methods we have presented thus far can be used on two-person zero-sum games of size 3 × 3 and larger. Specifically, when a saddle point is found in a game regardless of its size that saddle point will generate the solution to the game. The solution method of dominance will also be effective in reducing the size of large games to a point where perhaps one of the other methods we have discussed can be successfully applied. In this chapter we will illustrate examples of solutions to large games using saddle points and dominance. When these two methods cannot be applied with success, linear programming as a solution method may allow us to solve for the optimum strategies and the value of a game. We shall demonstrate later in this chapter just how to apply linear programming as a solution method for larger games.

SOLUTION BY SADDLE POINT

In the solution to 3 × 3 and larger games the first step, as in smaller games is to check for a saddle point, for if one exists it will make more difficult solution methods unnecessary. In Fig. 4-1 we have illustrated several games; in those where saddle points exist, we have shown the appropriate strategies for each of the players and the value of the game.

SOLUTION BY DOMINANCE

When one cannot find a saddle point in the original games, it is often possible to reduce the size of the game by eliminating rows or columns using the principles of dominance developed in earlier

$$X \begin{bmatrix} ② & 3 & 6 \\ -1 & 7 & 4 \\ -6 & 1 & 0 \end{bmatrix}$$

The payoff 2 is a saddle point. It is both the minimum value in its row and the maximum value in its column. Player Y will play his first column all of the time, and player X will counter by playing his first row all of the time.

$$X \begin{bmatrix} 3 & ㊀1 & 4 & 2 \\ -1 & -3 & -7 & 0 \\ 4 & -6 & 2 & -9 \end{bmatrix}$$

The payoff -1 is a saddle point; it is the smallest value in its row and the largest value in its column. Player Y will play his second column on each play and player X will respond by playing his first row on each play.

$$X \begin{bmatrix} 2 & -1 & 3 \\ 1 & 4 & 6 \\ -2 & 6 & 0 \\ 5 & 3 & -4 \end{bmatrix}$$

There is no payoff in this game which is both a row minimum and a column maximum. This game has no saddle point and cannot be solved by the saddle-point method.

$$X \begin{bmatrix} 4 & 2 & 1 & ⓪ & 7 \\ 6 & 5 & 2 & -1 & -3 \\ 3 & 0 & -6 & -1 & 4 \\ -8 & 7 & 4 & -6 & 3 \end{bmatrix}$$

The payoff zero is the smallest value in the first row and the largest value in the fourth column; both players will play strategies so as to make this value the value of the game. Y will play his fourth column on each play and X will counter by playing his first row. The game ends in a draw since the saddle-point value is zero.

$$X \begin{bmatrix} 2 & 1 & -9 \\ -6 & 2 & -11 \\ 1 & 7 & 7 \end{bmatrix}$$

There is no payoff in this game that is both the row minimum and the column maximum. This game does not contain a saddle point, and thus some other solution method must be employed.

Fig. 4-1. Illustration of saddle points.

chapters. Often when we employ this method the smaller game which results may contain a saddle point. It is also possible that the smaller game can be solved by using the algebraic method first developed in Chapter 2, or the method of subgames developed in Chapter 3.

Figure 4-2 illustrates a 3 × 3 game without a saddle point. Since the second column contains only positive payoffs and thus offers player X a guaranteed chance to win, player Y may not play the second column. Specifically, Y prefers column one to column two regardless of the choice of row by his opponent—that is, Y prefers a payoff of - 3 to 5; Y prefers to lose 1 point rather than 3 points;

Fig. 4-2. Game il-
lustrating domi-
nance.

$$X \begin{bmatrix} -3 & -2 \\ 1 & 7 \\ 4 & 2 \end{bmatrix}$$

Fig. 4-3.   Re-
duced   game
from Fig. 4-2.

with respect to the third row Y is indifferent between the equal
values 4.   The reduced game of size 3 × 2 appears in Fig. 4-3.   In
this reduced game, player X observes that playing his first row guarantees Y a chance to win and thus elects not to play this row.   The
game is now reduced to size 2 × 2.   The resulting 2 × 2 game does
not contain a saddle point, and thus must be solved by one of the
methods introduced in Chapter 2.   The strategies for each of the
players and the value of this game appear in Fig. 4-4.   If we take the

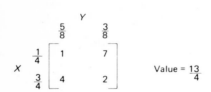

Fig. 4-4.   Reduced   2 × 2   game
with strategies and value.

Value = $\frac{13}{4}$

$$X \begin{array}{c} \\ 0 \\ \frac{1}{4} \\ \frac{3}{4} \end{array} \begin{bmatrix} -3 & 5 & -2 \\ 1 & 3 & 7 \\ 4 & 4 & 2 \end{bmatrix}$$

Fig. 4-5.   Strategies
shown   with   original
game.

strategies we have determined for the players and illustrate them using the original 3 × 3 game (we illustrate rows and columns not
played with zero strategies) we get the illustration shown in Fig. 4-5.

Figure 4-6a illustrates another game which can be reduced using
the principle of dominance.   In this case the original game is of size
4 × 3.   Examination of the original game reveals no saddle point;
therefore we next determine if dominance will be useful in reducing
the size of this game.   Player X observes that he prefers the payoffs
in row 2 to those in row 1 regardless of Y's choice of columns (- 4 is
preferable to - 8, 4 is preferred to - 6, and - 3 is a better alternative
than - 5).   For this reason, X will never play row 1 of the original
game.

The reduced game of size 3 × 3 appears in Fig. 4-6b.   There is
no saddle point in the reduced game.   Examination of this game re-

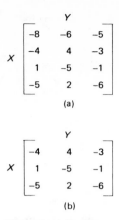

(a)

(b)

Fig. 4-6. (a) Exam-
ple of solution by
d o m i n a n c e, (b)
3 X 3 game result-
ing from part (a).

veals that X prefers the first row to the third row of this game regardless of Y's choice of columns (-4 is preferable to -5, 4 is preferred to
2, and -3 is a better alternative for X than -6). For this reason,
player X will never play row 3 of the game in Fig. 4-6b. The 2 X 3
game which results from this last application of the principle of
dominance appears in Fig. 4-7.

Fig. 4-7.   2 X 3
game resulting
from Fig. 4-6b.

Examination of the 2 X 3 game in Fig. 4-7 reveals that there is no
further possibility for application of dominance to reduce the game.
The game in its present form can be solved using the solution
methods developed in Chapter 3 for 2 X m games. We must find
solutions for each of the three possible 2 X 2 subgames, and then
choose as the game which would be played, the one with the greatest
possible benefit to player Y, i.e., the game with the lowest algebraic
value.

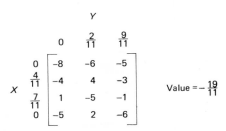

Fig. 4-8. Solution to three subgames.

In Fig. 4-8, we have illustrated the solution to and the values for each of the three possible subgames. It is obvious that player Y would choose to play subgame (c) from Fig. 4-8, with an expected value to him of $-19/11$ points per play. The strategies we have determined for the 2×2 subgame which would be played are illustrated on the original 4×3 game in Fig. 4-9, rows and columns not played being noted with zero strategies.

$$
\begin{array}{c}
& & Y \\
& & 0 \quad \tfrac{2}{11} \quad \tfrac{9}{11} \\
X & \begin{array}{c} 0 \\ \tfrac{4}{11} \\ \tfrac{7}{11} \\ 0 \end{array} & \left[\begin{array}{ccc} -8 & -6 & -5 \\ -4 & 4 & -3 \\ 1 & -5 & -1 \\ -5 & 2 & -6 \end{array}\right] & \text{Value} = -\tfrac{19}{11}
\end{array}
$$

Fig. 4-9. Strategies illustrated
on original game.

GAMES THAT CANNOT BE REDUCED BELOW 3 X 3

When there is no saddle point in a game, and when reduction of the game by dominance is not possible, linear programming offers an efficient solution method. Those readers familiar with the principles of linear programming recognize it as a technique for maximizing or minimizing some objective function subject to certain constraints. Both the objective function and the constraints must be stated in the form of linear equations in order for this technique to be used. It is possible to generate such objective functions and constraints in games using the inequalities representing the guarantees to the game participants first introduced in Chapter 3.

The game illustrated in Fig. 4-10 does not contain a saddle point

$$X \begin{bmatrix} 1 & 2 & -1 \\ -2 & 1 & 1 \\ 2 & 0 & 1 \end{bmatrix}$$

Y

Fig. 4-10. Game illustrating solution by linear programming.

$$\begin{array}{c} & Y_1 \quad Y_1 \quad Y_1 \\ X_1 \\ X_2 \\ X_3 \end{array} \begin{bmatrix} 1 & 2 & -1 \\ -2 & 1 & 1 \\ 2 & 0 & 1 \end{bmatrix}$$

Fig. 4-11. Illustration of notation.

and cannot be reduced to a smaller game by the use of the principle of dominance. In a situation such as this we must turn to linear programming for the determination of the optimum strategies and the game value. In the game in Fig. 4-10, we are seeking to determine just what proportion of the time each player should spend playing his respective rows and columns. In Fig. 4-11, we have designated these six unknowns with appropriate subscript notation. Our linear programming solution to this game must therefore yield the answers for the six unknowns X_1, X_2, X_3, Y_1, Y_2, and Y_3.

Beginning first with the determination of player Y's optimum strategies, we can write his expectations using the methods introduced in Chapter 3 as follows:

$$1Y_1 + 2Y_2 - 1Y_3 \leqslant V$$
$$-2Y_1 + 1Y_2 + 1Y_3 \leqslant V$$
$$2Y_1 + 0Y_2 + 1Y_3 \leqslant V$$

where V = the value of the game. We interpret these three inequalities as meaning the player Y expects to win the value of the game or less (a more negative average payoff), regardless of X's choice of row strategies; this assumes we correctly determine Y_1, Y_2, and Y_3.

To remove the V terms from the right-hand side of the three inequalities, we can divide all of the terms by V.

$$\frac{Y_1}{V} + \frac{2Y_2}{V} - \frac{Y_3}{V} \leqslant 1$$
$$\frac{-2Y_1}{V} + \frac{Y_2}{V} + \frac{Y_3}{V} \leqslant 1$$

$$\frac{2Y_1}{V} + \frac{0Y_2}{V} + \frac{Y_3}{V} \leqslant 1$$

to remove the V terms in the denominators of the Y terms, let us define a new Y variable.

$$\overline{Y} = \frac{Y}{V}$$

and then solve the game in terms of $\overline{Y}$'s instead of Y's. Then when we have found a solution in terms of $\overline{Y}$'s, we can multiply each of our $\overline{Y}$'s by V to get the appropriate Y's. (If $\overline{Y} = Y/V$, then of course $Y = \overline{Y} \cdot V$). This is simply a bit of helpful mathematical juggling. Our three inequalities thus become:

$$\overline{Y}_1 + 2\overline{Y}_2 - \overline{Y}_3 \leqslant 1$$
$$-2\overline{Y}_1 + \overline{Y}_2 + \overline{Y}_3 \leqslant 1$$
$$2\overline{Y}_1 + 0\overline{Y}_2 + \overline{Y}_3 \leqslant 1$$

We realize that the values we finally determine for Y_1, Y_2, and Y_3 must sum to unity; this requirement becomes another constraint under which we must solve the game. It is written in symbolic form as:

$$Y_1 + Y_2 + Y_3 = 1$$

To put this latest equation into the $\overline{Y}$ form like the three previous inequalities, we once again divide all its terms by V.

$$\frac{Y_1}{V} + \frac{Y_2}{V} + \frac{Y_3}{V} = \frac{1}{V}$$

and then letting $\overline{Y} = Y/V$ as we did above, we finally get

$$\overline{Y}_1 + \overline{Y}_2 + \overline{Y}_3 = \frac{1}{V}$$

Repeating all four constraints under which we must now solve the game, we have

$$\overline{Y}_1 + \overline{Y}_2 + \overline{Y}_3 = \frac{1}{V}$$
$$\overline{Y}_1 + 2\overline{Y}_2 - \overline{Y}_3 \leqslant 1$$
$$-2\overline{Y}_1 + \overline{Y}_2 + \overline{Y}_3 \leqslant 1$$
$$2\overline{Y}_1 + 0\overline{Y}_2 + \overline{Y}_3 \leqslant 1$$

Y's objective in any game is to *minimize* the value of V; this is identical to *maximizing* $1/V$. Thus player Y's objective is to maximize the value of $\overline{Y}_1 + \overline{Y}_2 + \overline{Y}_3$, subject to three linear constraints. In the more usual linear programming notation, the problem would be stated as illustrated in Fig. 4-12.

Maximize:

$$\overline{Y}_1 + \overline{Y}_2 + \overline{Y}_3$$

subject to:

$$\overline{Y}_1 + 2\overline{Y}_2 - \overline{Y}_3 \leqslant 1$$
$$-2\overline{Y}_1 + \overline{Y}_2 + \overline{Y}_3 \leqslant 1$$
$$2\overline{Y}_1 + 0\overline{Y}_2 + \overline{Y}_3 \leqslant 1$$

Fig. 4-12. Formal statement of the problem.

Those familiar with linear programming will recognize these three inequalities as being of type I; in this case we need to add a slack variable to each of the inequalities to produce equations for the simplex method of linear programming. In Fig. 4-13, we illustrate the problem ready for insertion into the first simplex table. Slack variables have been denoted as S_1, S_2, and S_3 in this case.

Maximize:

$$\overline{Y}_1 + \overline{Y}_2 + \overline{Y}_3$$

subject to:

$$\overline{Y}_1 + 2\overline{Y}_2 - \overline{Y}_3 + S_1 + 0S_2 + 0S_3 = 1$$
$$-2\overline{Y}_1 + \overline{Y}_2 + \overline{Y}_3 + 0S_1 + S_2 + 0S_3 = 1$$
$$2\overline{Y}_1 + 0\overline{Y}_2 + \overline{Y}_3 + 0S_1 + 0S_2 + S_3 = 1$$

Fig. 4-13. Equations for the first simplex table.

Three iterations are required for the solution to this problem. These are illustrated in Fig. 4-14. The simplex procedure and notation used are standard. Details of the calculations have been omitted.

Examination of the final simplex table reveals that the value of the objective function is $17/11$. We remember this to be $1/V$; thus the value of this game is $11/17$. Since we know the value of the game, we can now convert our $\overline{Y}$ terms into Y terms by multiplying each of

First Simplex Table

C_j			1	1	1	0	0	0	
	Basis	P_0	$\bar{Y}_1$	$\bar{Y}_2$	$\bar{Y}_3$	S_1	S_2	S_3	
0	S_1	1	1	2	−1	1	0	0	$\theta = 1$
0	S_2	1	−2	1	1	0	1	0	$\theta = -1/2$
0	S_3	1	2	0	1	0	0	1	$\theta = 1/2$ (this row is replaced)
	Z_j	0	0	0	0	0	0	0	
	$C_j - Z_j$		1	1	1	0	0	0	

Since there is a tie in the $C_j - Z_j$ row, we have arbitrarily let $\bar{Y}_1$ be the first entering variable.

Second Simplex Table

C_j			1	1	1	0	0	0	
	Basis	P_0	$\bar{Y}_1$	$\bar{Y}_2$	$\bar{Y}_3$	S_1	S_2	S_3	
0	S_1	1/2	0	2	−3/2	1	0	−1/2	$\theta = 1/4$ (this row is replaced)
0	S_2	2	0	1	2	0	1	1	$\theta = 2$
1	$\bar{Y}_1$	1/2	1	0	1/2	0	0	1/2	$\theta = $ not defined
	Z_j	1/2	1	0	1/2	0	0	1/2	
	$C_j - Z_j$		0	1	1/2	0	0	−1/2	

$\bar{Y}_2$ enters next.

Third Simplex Table

C_j			1	1	1	0	0	0	
	Basis	P_0	$\bar{Y}_1$	$\bar{Y}_2$	$\bar{Y}_3$	S_1	S_2	S_3	
1	$\bar{Y}_2$	1/4	0	1	−3/4	1/2	0	−1/4	$\theta = -1/3$
0	S_2	7/4	0	0	11/4	−1/2	1	5/4	$\theta = 7/11$ (this row is replaced)
1	$\bar{Y}_1$	1/2	1	0	1/2	0	0	1/2	$\theta = 1$
	Z_j	3/4	1	1	−1/4	1/2	0	1/4	
	$C_j - Z_j$	0	0	0	5/4	−1/2	0	−1/4	

$\bar{Y}_3$ enters next.

Final Simplex Table

C_j			1	1	1	0	0	0
	Basis	P_0	$\bar{Y}_1$	$\bar{Y}_2$	$\bar{Y}_3$	S_1	S_2	S_3
1	$\bar{Y}_2$	8/11	0	1	0	4/11	3/11	1/11
1	$\bar{Y}_3$	7/11	0	0	1	−2/11	4/11	5/11
1	$\bar{Y}_1$	2/11	1	0	0	1/11	−2/11	3/11
	Z_j	17/11	1	1	1	3/11	5/11	9/11
	$C_j - Z_j$		0	0	0	−3/11	−5/11	−9/11

(Since there is no positive entry in the $C_j - Z_j$ row we have reached the optimum answer.)

Fig. 4-14. Simplex solution for Y's strategies.

them by V as follows:

$$Y_1 = \bar{Y}_1 \cdot V$$
$$Y_2 = \bar{Y}_2 \cdot V$$
$$Y_3 = \bar{Y}_3 \cdot V$$

$$Y_1 = \frac{2}{11} \cdot \frac{11}{17} \text{ or } \frac{2}{17}$$

$$Y_2 = \frac{8}{11} \cdot \frac{11}{17} \text{ or } \frac{8}{17}$$

$$Y_3 = \frac{7}{11} \cdot \frac{11}{17} \text{ or } \frac{7}{17}$$

Total $= {}^{17}/_{17}$ or 1

Having determined Y's optimum strategies, we now turn our attention to player X. The most efficient method for finding player X's strategies is to read them directly from the final simplex table of Fig. 4-14. They appear there as the entries in the $C_j - Z_j$ row under the slack variable columns, i.e., $-{}^3/_{11}$, $-{}^5/_{11}$, and $-{}^9/_{11}$. Like $\bar{Y}$, however, these three values are in fact $\bar{X}$ values and must each be multiplied by V to produce the appropriate X values; this calculation is

$$X_1 = \bar{X}_1 \cdot V$$
$$X_2 = \bar{X}_2 \cdot V$$
$$X_3 = \bar{X}_3 \cdot V$$

$$X_1 = \frac{3}{11} \cdot \frac{11}{17} \text{ or } \frac{3}{17}$$

$$X_2 = \frac{5}{11} \cdot \frac{11}{17} \text{ or } \frac{5}{17}$$

$$X_3 = \frac{9}{11} \cdot \frac{11}{17} \text{ or } \frac{9}{17}$$

Total ${}^{17}/_{17}$ or 1

We disregard the algebraic sign, since of course negative values for strategies would have no real meaning to the players. The student of linear programming will recognize this method as employing the concept of the dual.

It is possible to produce X's optimum strategies directly by linear programming just as we were able to solve the game from the point of view of player Y. We first set up the inequalities which represent the expectations of X as follows.

$$1X_1 - 2X_2 + 2X_3 \geqslant V$$
$$2X_1 + 1X_2 + 0X_3 \geqslant V$$
$$-X_1 + 1X_2 + 1X_3 \geqslant V$$

Dividing through by V we get

$$\frac{X_1}{V} - \frac{2X_2}{V} + \frac{2X_3}{V} \geqslant 1$$

$$\frac{2X_1}{V} + \frac{X_2}{V} + \frac{0X_3}{V} \geqslant 1$$

$$\frac{-X_1}{V} + \frac{X_2}{V} + \frac{X_3}{V} \geqslant 1$$

Then defining $\bar{X}$ as being equal to X/V we produce

$$\bar{X}_1 - 2\bar{X}_2 + 2\bar{X}_3 \geqslant 1$$
$$2\bar{X}_1 + \bar{X}_2 + 0\bar{X}_3 \geqslant 1$$
$$-\bar{X}_1 + \bar{X}_2 + \bar{X}_3 \geqslant 1$$

Player X desires to *maximize* V which is the same as *minimizing* $1/V$; since $X_1 + X_2 + X_3$ must equal 1, $\bar{X}_1 + \bar{X}_2 + \bar{X}_3$ must equal $1/V$; thus our linear programming problem becomes

minimize:

$$\bar{X}_1 + \bar{X}_2 + \bar{X}_3$$

subject to:

$$\bar{X}_1 - 2\bar{X}_2 + 2\bar{X}_3 \geqslant 1$$
$$2\bar{X}_1 + \bar{X}_2 + 0\bar{X}_3 \geqslant 1$$
$$-\bar{X}_1 + \bar{X}_2 + \bar{X}_3 \geqslant 1$$

Since the inequalities just written are of type II, we will first have to subtract a slack variable S from each inequality to produce an

Minimize:

$$\bar{X}_1 + \bar{X}_2 + \bar{X}_3$$

subject to:

$$\bar{X}_1 - 2\bar{X}_2 + 2\bar{X}_3 - S_1 + 0S_2 + 0S_3 + A_1 + 0A_2 + 0A_3$$
$$2\bar{X}_1 + \bar{X}_2 + 0\bar{X}_3 + 0S_1 - S_2 + 0S_3 + 0A_1 + A_2 + 0A_3$$
$$-\bar{X}_1 + \bar{X}_2 + \bar{X}_3 + 0S_1 + 0S_2 - S_3 + 0A_1 + 0A_2 + A_3$$

Fig. 4-15. Equations for simplex solution for X's strategies.

equation, and then add an artificial variable A to each equation to produce a feasible beginning basis. The completed set of equations ready for the first simplex minimizing table appears in Fig. 4-15; we have omitted the simplex tables in this instance, since the procedure is the same as for maximizing except that the simplex criterion for evaluation of the entering variable in this case would be the most negative $C_j - Z_j$ value.

Linear programming offers an efficient method for determining the strategies and value of larger games where the solution methods introduced earlier cannot be applied. Whereas the size of the simplex table to be iterated would become quite large for some games, there are computer programs which solve these large linear programming problems quite easily.

CHAPTER 5

Nonzero-Sum Games

We have imposed two major restrictions on all of the games which have been discussed in the first four chapters:

1. Only two persons or opposing interests were involved.
2. The payoffs for both players were equal in magnitude but opposite in sign so that the sum of the payoffs for all pairs of choices was zero.

In this chapter we will consider games for which the second of these restrictions has been removed. Such games are generally called nonzero-sum games to emphasize the fact that the sum of the players' payoffs for some pairs of choices will differ from zero. In Chapter 6 we will examine games which involve more than two opposing interests. These games are called *n*-person games.

In a two person zero-sum game, one player's gain is always the other player's loss. Thus there is no reason to consider the possibility of cooperation or negotiation between the players. However, the existence of more than two players and/or payoffs that do not add to zero introduces the possibility of cooperation and bargaining. For example, in an *n*-person game two or more players may decide to cooperate in the hope that by acting together they can more easily beat the opposition. Similarly, when the sum of the payoffs is not zero the players may be able to cooperate in such a way that they will maximize the total payoff rather than maximizing the payoff to a single player.

Although negotiation is possible in *n*-person and nonzero-sum games it is *not required*. Therefore, it is convenient to divide the analysis of these games into two parts: games in which the participants can negotiate, and games in which negotiation is not permitted. In Chapters 5 and 6 we will assume that the players can negotiate, bargain, and cooperate. In Chapter 7 we will examine noncooperative or nonnegotiable games.

PAYOFF MATRIX FOR A NONZERO-SUM GAME

The payoffs for zero-sum games have been presented in a matrix which actually shows only half of the payoffs. By convention, the matrix shows the payoffs for player X; the payoffs for player Y are obtained by changing the sign of each payoff.

For nonzero-sum games it is necessary to explicitly show the payoffs for both players for each possible combination of choices. This can be accomplished either by using a simple payoff matrix for each player or by using a single matrix which simultaneously displays the payoffs for both players.

To illustrate the mechanics of recording the payoffs in a nonzero-sum game, consider the following matching game. Player X and player Y both put one hand behind their back. On the count of three, each player simultaneously reveals his hand, holding up one to five fingers. The payoffs, which depend on whether the number of fingers that is held up by each player is even or odd, are shown in Fig. 5-1.

Choice		Payoff	
Player X	Player Y	Player X	Player Y
Even	Even	1	-2
Even	Odd	-2	2
Odd	Even	0	5
Odd	Odd	3	1

Fig. 5-1. Payoffs for a nonzero-sum game.

In Fig. 5-2, the payoffs for this game are recorded in two matrices which are in the same form as payoff matrices for zero sum games.

		Y's Choices				Y's Choices	
		Even	Odd			Even	Odd
	Even	1	-2		Even	-2	2
X's Choices				X's Choices			
	Odd	0	3		Odd	5	1

(a) Payoffs for Player X   (b) Payoffs for Player Y

Fig. 5-2. Payoff matrices for a nonzero-sum game.

It is usually more convenient to record both sets of payoffs for a nonzero-sum game in a single matrix in which pairs of payoffs are recorded for each possible combination of choices. The first element in each pair of payoffs is the payoff for the player whose choices are written to the left of the matrix (usually player X). The second element represents the payoff for player Y whose choices are written above the matrix. This combined form of the payoff matrix is illustrated in Fig. 5-3.

	Y's Choices	
	Even	Odd
Even	1,-2	-2,2
Odd	0, 5	3,1

X's Choices

Fig. 5-3. Combined payoff matrix
for a nonzero-sum game.

Payoffs for nonzero-sum games are usually recorded in the combined form. However, for purposes of analysis it is often convenient to record the payoffs in two separate matrices.

THE GAME DIAGRAM

The payoffs for a two-person nonzero-sum game can also be represented in a two-dimensional diagram in which the horizontal axis represents the payoffs for player X and the vertical axis the payoffs for player Y. The diagram for the finger matching game discussed above is shown in Fig. 5-4.

The quadrangle $RSTU$, which is formed by connecting the points which represent the four possible outcomes for a single play of the game by straight lines, is called the *payoff polygon* for the game. All possible pairs of average payoffs which the players can attain by repeated play of the game are represented by points inside or on the border of this payoff polygon. For example, if the probability of displaying an even number of fingers is $1/2$ for both players and if the choices of even or odd are randomly distributed, then each of the four possible outcomes will occur with a probability of $1/4$, in the long run, and the average pair of payoffs for the players will be $(1/2, 1^1/2)$. It should be noted that this pair of payoffs, which is point A in Fig. 5-4, is inside the payoff polygon.

As defined in Chapter 2, the value of a two-person zero-sum game is the average payoff that player X can expect to win for each play of the game. The average payoff for player Y is not computed because it is simply the negative of the average payoff for player X. In a nonzero-sum game this simple relationship between the average payoffs for the two players is no longer valid so it is necessary to compute the value of the game for each player. The average payoff for each player is computed in exactly the same way as the value of a

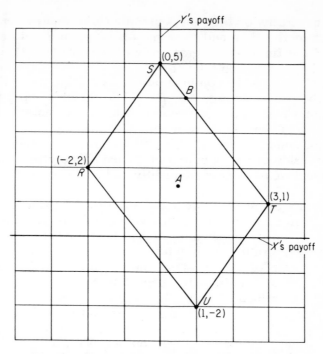

Fig. 5-4. Game diagram for finger-matching game.

zero-sum game. The computation of the pair of average values for the above example is illustrated in Fig. 5-5.

If the players make their choices in such a way that only two of the four possible outcomes occur, the pair of average payoffs will lie on the line connecting the two points which represent the two outcomes which occur. The exact location on the line will depend on the relative frequencies with which the two outcomes occur. Thus if player X always chooses an odd number of fingers and player Y chooses an even number of fingers 75 percent of the time and an odd

Choice		Probability	Payoff		Expected Payoff	
X	Y	of Occurrence	X	Y	X	Y
E	E	¼	1	−2	¼	−½
E	O	¼	−2	2	−½	½
O	E	¼	0	5	0	1¼
O	O	¼	3	1	¾	½
					½	1½

Fig. 5-5. Computation of the average payoff for both players of a nonzero-sum game.

number 25 percent of the time, outcome (0,5), point S, will occur 75 percent of the time and outcome (3, 1), point T, will occur 25 percent of the time. The average pair of payoffs (3/4, 4), which is point B in Fig. 5-4, lies on the line ST.

It is not possible to combine outcomes for this game in such a way that the pair of average payoffs will lie outside the payoff polygon without violating the basic constraints that are imposed on the probabilities of choosing even or odd; namely, P (even) $\geqslant 0$, P (odd) $\geqslant 0$, and P (even) + P (odd) = 1. Thus, all possible pairs of payoffs lie inside or on the border of the payoff polygon. The set of all such pairs of payoffs is frequently called the *payoff space* for the game.

The results obtained above for one particular game can be extended to other games providing the geometric figure which is formed when the outcomes are joined by straight lines is a convex polygon.[1] The generalized results, which apply to any game which has at least three distinct pairs of payoffs as outcomes, can be stated as follows.

If *some* of the points in the payoff space are joined to form a convex polygon which includes the payoffs for all outcomes either on its boundaries or inside, then the convex polygon will also include all of the payoffs, and only those payoffs, which can be realized from all possible combinations of the outcomes for the game. Such a polygon is called the *convex hull* of the set of payoff points. Figure 5-6 illustrates an incorrect and correct payoff polygon for a nonzero-sum game which has the following payoff matrix.

$$Y$$

		Y_1	Y_2	Y_3
	X_1	0,5	-2,2	3,3
X	X_2	1,-2	3,1	1,1

Figure 5-6a, which was constructed by joining *all* pairs of payoffs, is not a convex polygon because a line can be drawn from one point in the polygon to another point in the polygon such that all the points on the line are not included in or on the boundary of the polygon. One such line, drawn from point ($1/2$, -$1/2$) to (2, 2), is shown in Fig. 5-6a. Alternatively, if the players make their decisions so the decision pairs (X_2, Y_1) and (X_2, Y_2) each occur 50 percent of the time, the average pair of payoffs will be (2, -$1/2$), a point which is not inside the polygon drawn in Fig. 5-6a. In addition, the interior

[1] A convex polygon has the following properties: (1) if two points inside or on the boundary of the polygon are joined by a straight line, every point on that line lies inside or on the boundary of the polygon, and (2) all interior angles of the polygon are less than 180 degrees.

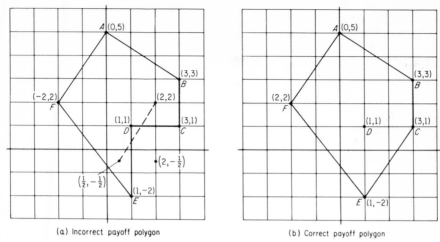

(a) Incorrect payoff polygon   (b) Correct payoff polygon

Fig. 5-6. Incorrect and correct payoff polygons for a nonzero-sum game.

angle formed by the lines CD and DE is greater than $180°$. The correct payoff polygon is shown in Fig. 5-6b. This polygon, which is the convex hull for the game, connects only five outcome pairs, but the sixth pair of payoffs, $(1, 1)$, is an interior point of the polygon.

It should be noted that several different games may have exactly the same payoff polygon. For example, the payoff for decision pair (X_2, Y_3) for the game shown in Fig. 5-6 could be changed to $(-1, +1)$ without changing the payoff polygon. In fact, if fractional payoffs are permitted, there are an infinite number of games which have the same payoff polygon as drawn in Fig. 5-6 because there are an infinite number of different payoffs for the decision pair (X_2, Y_3) which are inside the payoff polygon.

Even more important is the fact that the payoff polygon can be exactly the same for games which have different solutions. Two such games and their payoff polygon are shown in Fig. 5-7.

In Game 1, player X should always choose X_1 because his payoff for X_1 is greater than his payoff for X_2 regardless of the choice made by player Y. (X_1 is a dominate strategy for player X). Player Y is in a similar situation; he should always choose Y_2 to maximize his payoff. Thus the solution for this game is the decision pair (X_1, Y_2) and the payoff is $(4, 6)$.

In Game 2, X_1 is a dominate strategy for player X and Y_1 is a dominate strategy for player Y. Therefore, the decision pair (X_1, Y_1) with payoff $(5, 5)$ is the solution for this game. Although the payoff polygon is the same for both games, the solution is different. Thus the construction of a game diagram, by itself, is not sufficient to determine the solution for a nonzero-sum game.

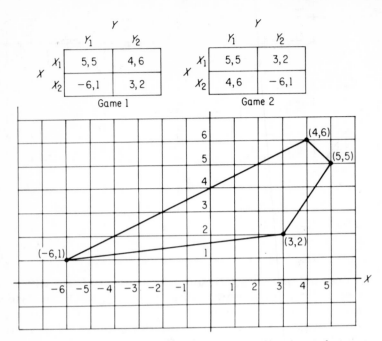

Fig. 5-7. Two games with the same payoff polygon but different solutions.

NEGOTIATIONS

There is no basis for negotiation in a zero-sum game because one player's gain is always the other player's loss. However, in a nonzero-sum game a shift from one solution to another may result in a gain for both players so the possibility of negotiation must be considered.

In general, conflicting parties enter into negotiations if each player feels he can gain from the results. Since game theory does not consider differences in the player's ability to negotiate, the desirability of initiating negotiations and the results of these negotiations depend entirely on the characteristics of the payoff matrix.

If the interests of the conflicting players completely coincide, the solution of the negotiated game is trivial. As an example of such a game, consider the finger-matching game which has the payoffs shown in Fig. 5-8. If both players show the same number of fingers they both receive one dollar for each finger displayed; if they display a different number of fingers each player must pay the house one dollar for each finger displayed.

If the players are able to negotiate with each other and if their decisions are not restricted in any way, they will both agree to hold

Number of Fingers	Y				
	1	2	3	4	5
1	1, 1	-1,-2	-1,-3	-1,-4	-1,-5
2	-2,-1	2, 2	-2,-3	-2,-4	-2,-5
X 3	-3,-1	-3,-2	3, 3	-3,-4	-3,-5
4	-4,-1	-4,-2	-4,-3	4, 4	-4,-5
5	-5,-1	-5,-2	-5,-3	-5,-4	5, 5

Fig. 5-8. A nonzero-sum game in which
the players have coinciding interests.

up five fingers on each play of the game and thus receive an average payoff of (5,5). The solutions of nonzero-sum games in which the players have completely coinciding interests are of little interest.

The interesting features of nonzero-sum games occur whenever the payoff matrix is such that the interests of the players partly conflict and partly coincide. It is then possible for the players to enter into negotiations in an attempt to use their common interests as leverage to bring about a settlement of their conflicting interests. An example of such a game is shown in Fig. 5-9.

	Y	
	Y_1	Y_2
X X_1	(5,5)	(16,2)
X_2	(2,16)	(6,6)

Fig. 5-9. A negotiable nonzero-sum game.

The first choice for each player dominates his second choice. Thus in the absence of negotiations we would expect the solution to be decision pair (X_1, Y_1) with a payoff of (5,5). Neither player would be willing to select his second choice by himself because he would reduce his payoff from 5 to 2 and increase his competitor's payoff from 5 to 16. However, if the players could communicate they could both agree to always select their second choice and thus increase the payoff for both players from 5 to 6. Furthermore, if both players agreed to coordinate their choices so the decision pairs (X_1, Y_2) and (X_2, Y_1) each occurred half of the time, the players could increase their average payoff to (9,9). It should be noted that the latter result will not be obtained if each player selects his first and second choice by some random device such as tossing a coin. If such a random selection procedure is used, each of the four outcomes would occur one-quarter of the time and the average payoff would be $(7^1/_4, 7^1/_4)$. Coordinated choices required an agreement on

both the frequency with which the choices will be made and the se-
quence in which they will be made.

The negotiations for the game in Fig. 5-9 are relatively simple
because neither player has a bargaining advantage. The negotiations
are more complicated when one player has a bargaining advantage
such as in the game shown in Fig. 5-10.

$$Y$$

		Y_1	Y_2
X	X_1	(1,4)	(-1,-4)
	X_2	(-3,-1)	(4,1)

Fig. 5-10. A game in which player
Y has a bargaining advantage.

In this game player Y would prefer the solution (X_1,Y_1) and
player X would prefer the solution (X_2,Y_2). Neither player has a
dominate strategy nor a safe strategy (i.e., a strategy which will
guarantee him a positive payoff regardless of the choice made by his
competitor). In such a situation it might seem reasonable for the
players to agree to coordinate their choices so the decision pairs
(X_1,Y_1) and (X_2,Y_2) each occur half of the time thus giving the
players an average payoff of $(2^1/_2, 2^1/_2)$. However, if player Y
threatens to always select choice Y_1 there is little that Player X can
do against him. If player X chooses X_1 player Y will receive a payoff
of 4 and X will receive a payoff of only 1. If X wants to "punish"
player Y he can choose X_2 thus inflicting a loss of -1 on his competitor but X must be willing to accept a payoff of -3, a loss that is
three times as great as the loss inflicted on Y. Thus player Y has a
threat strategy which he can use to gain a bargaining advantage.

Player X does not have a similar threat strategy. If he always
plays X_1, Y will choose Y_1 and receive a payoff that is four times
greater than the payoff for player X. If X threatens to choose X_2 his
competitor can select Y_1; both players receive a negative payoff but
X again suffers a loss that is three times greater than the loss of
player Y, so player X's threat is very weak.

The selection of a procedure for determining negotiated solutions
for nonzero-sum games which adequately reflect the bargaining advantages and threat potentials of the participants is not obvious. In
the remainder of this chapter we will examine several theories of
bargaining and arbitration which have been employed in attempting
to solve such negotiation problems.

THE SOLUTION OF NEGOTIATED GAMES

To find a solution for a negotiated game we first define a set of acceptable solutions, usually called the *negotiation set*, and then select one particular solution from the negotiation set. The procedure for defining the negotiation set is derived from the logical considerations of game theory. It is assumed that the players are intelligent, rational individuals and that each player is fully informed on the structure of the game and the payoffs for both players. The selection of a particular solution from the negotiation set is based primarily on psychological considerations. The particular assumptions made are reflected in the selection of a *status quo point* (defined below) which is a major determinant of the final solution. Since it is possible to introduce a large number of different (but reasonable) psychological assumptions, a given game may have several status quo points. There are also several different procedures for combining the negotiation set and the status quo point to obtain a final solution. Thus each game usually has several solutions for each status quo point.

We will examine the steps for obtaining a solution to a nonzero-sum game in the order in which they are usually performed: (1) the determination of the negotiation set, (2) the selection of a status quo point and (3) the combination of the status quo point and the negotiation set to obtain a solution for a particular game.

THE DETERMINATION OF THE NEGOTIATION SET

For each player in a nonzero-sum game there exists a mixed strategy which the player can use to guarantee that he will obtain some minimum average payoff regardless of the strategy which is used by his opponent. We shall call the value of this average payoff the player's *security level*. Because a player can be sure of receiving a payoff equal to this security level without the cooperation of his opponent, the security level represents the lowest payoff a rational player would accept from a negotiated solution for a game.

Security levels are determined by assuming that the interests of the opposing player are diametrically opposed to one's own interests. Since this is the basic assumption made in all zero-sum games, the security levels for the players of a nonzero-sum game are determined by obtaining the minimax solution for the two zero-sum games which are formed from the payoffs for each player in the nonzero-sum game. As an example we will consider the finger-matching game which was introduced at the beginning of this chapter. The payoff matrix for

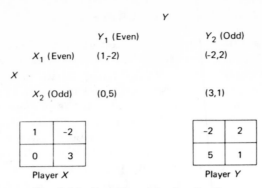

Fig. 5-11. Payoff matrix for the finger matching game and the associated zero-sum games for each player.

this game and the two associated zero-sum games which contain the payoffs for each player are shown in Fig. 5-11.

Using the techniques illustrated in the first four chapters we obtain the minimax solution for each of the zero-sum games shown in Fig. 5-11. The minimax solution for the zero-sum game consisting of the payoffs for player X is the mixed strategy $(1/2, 1/2)$; i.e., play "even" 50 percent of the time and "odd" 50 percent of the time. If player Y chooses "even" with probability p, thus playing "odd" with probability $(1 - p)$, the payoff for player X will be

$$\text{Payoff } (X) = 1/2\, p\, (1) + 1/2\, p\, (0) + 1/2\, (1 - p)\, (-2) + 1/2\, (1 - p)\, (3)$$
$$= 1/2\, p \quad\quad + 0 \quad\quad - 1 \quad + p \quad\quad +1 1/2 \quad - 1 1/2\, p$$
$$= 1/2$$

By using this minimax strategy, player X can assure himself an average payoff of $1/2$ without any cooperation from his competitor, thus the security level for player X is $1/2$.

Similarly, the minimax solution of the zero-sum game composed of the payoffs for player Y is the mixed strategy $(1/8, 7/8)$. This strategy will assure player Y of an average payoff of $1 1/2$ regardless of the choices made by player X; thus $1 1/2$ is the security level for player Y. Figure 5-12, which is a diagram of the matching game, shows the security levels for both players.

By extending a horizontal and vertical line from the point representing the security levels for both players (Point S), it is possible to define an area of feasible solutions; area SAB for the game presently under consideration. We assume that the players are rational and therefore would not accept any negotiated solution to the left of line SA or below line SB.

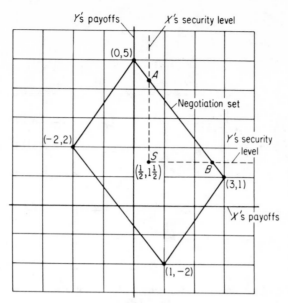

Fig. 5-12. Game diagram showing players security levels and the negotiation set.

Assuming the utility of player Y is not affected by increases in the payoff[2] to player X, Y would be indifferent between the solutions represented by the points on the line SB. Player X obviously has a strong preference for the solution at point B, since this solution would result in the highest possible payoff for X given a payoff of $1\frac{1}{2}$ for player Y. Similarly, player Y clearly prefers a solution at point A to any other solution on the line SA. In fact, given any solution inside or on the boundary of SAB, either player could increase his payoff without decreasing the payoff of his competitor except when the solution is on the line AB. Thus, rational players should not accept any negotiated solution unless it is on the line AB. The points on this line are the *negotiation set* for the game. Similar negotiation sets can be determined for any nonzero-sum game.

Sometimes one whole boundary of the payoff polygon will be included between the players' security levels. This situation is diagrammed in Fig. 5-13. The negotiation set for this game is the entire line AB. Points on the lines CA and DB are not in the negotiation set because it is possible for both players to simultaneously improve their payoffs by moving from points on either of these lines.

[2] This assumption implies that the players are primarily interested in absolute payoffs and not their share of the total payoff.

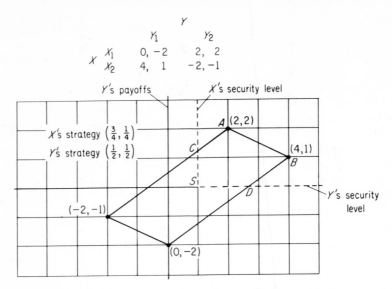

	Y	
	Y_1	Y_2
X_1	0, −2	2, 2
X_2	4, 1	−2, −1

Fig. 5-13. A game for which the negotiation set is the line AB.

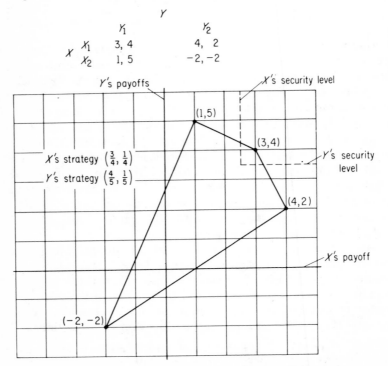

	Y	
	Y_1	Y_2
X_1	3, 4	4, 2
X_2	1, 5	−2, −2

Fig. 5-14. Game with a more complex negotiation set.

It is also possible for the negotiation set to include segments of two or more lines as illustrated in Fig. 5-14.

Many other configurations are possible, especially for larger games which may have more than two choices per player. Regardless of the geometric configuration, every point in the negotiation set must satisfy the following two conditions: (1) It should not be possible for both players to simultaneously improve their payoffs by moving from a point in the negotiation set, and (2) the average payoff for each player must not be smaller than his security level. These two conditions, together with the payoff polygon for the game, completely define the set of points that would be acceptable negotiated solutions for two rational players.

THE STATUS QUO POINT

The status quo point is a particular point inside or on the boundary of the payoff polygon which serves as a point of reference for the solution of a negotiated game. It is the point from which bargaining proceeds. Although any point in the payoff polygon can serve as the status quo point, the point which is selected must reflect differences in the players' bargaining power or threat potential if meaningful solutions are to be obtained.

One obvious difference between the players of a game is the difference in their security levels. Thus it has been proposed that the point which has the coordinates of the players' security levels would be a good status quo point. Since this point represents the payoffs which the players can guarantee to themselves without cooperation, it is certainly reasonable to argue that the bargaining should begin at this point. Furthermore, the entire set of payoffs is involved in the computation of the security levels so, it is argued, the status quo point should reflect all important differences in the players' strengths and weaknesses. However, some game theorists feel that the difference between players is not adequately measured by the difference in their security levels so they have proposed other methods for determining the location of the status quo point.

One difference which may be completely obscured if security levels are used as the status quo point is the difference in the players' threat potentials. Before we can determine the threat potentials for the players we must look at the process of making a threat in a game.

Game theory assumes that the players are rational persons and thus will not make a threat just for the sake of making a threat. Therefore, if a player exercises a threat he *must* carry it out if his opponent fails to comply with his demands. The process of making a threat follows a three-step bargaining scheme.

1. Player X announces a threat strategy, T_x.
2. Player Y, in ignorance of X's threat strategy, announces a threat strategy, T_y.
3. Players X and Y bargain. If they reach an agreement then that agreement becomes effective. If they do not come to an agreement they *must* use their threat strategies.

The question of how the players are forced to carry out their threats arises naturally but it is not relevant to game theory. We assume that the players are in some way required to carry out their threats and the payoffs for the players are determined in this manner.

The threat potential concept is illustrated by the games in Fig. 5-15.

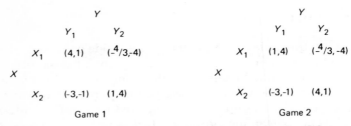

Fig. 5-15. Two games illustrating the threat-potential concept.

In Game 1 of Fig. 5-15, player X threatens player Y by convincing Y that he will always choose X_1 unless Y is willing to negotiate a solution. Player Y has three choices: (1) he can negotiate, (2) he can refuse to negotiate and select Y_1 thus earning a payoff of 1 as opposed to player X's payoff of 4, or (3) he can refuse to negotiate and select Y_2 thus losing three times as much as player X.

Player Y does not have as effective a threat strategy in Game 1. At best, Y can threaten X by convincing him that he will always choose Y_2 unless X is willing to comply with Y's demands. Faced with such a threat player X may be willing to negotiate to avoid a negative payoff, but there is a good possibility that he will simply choose X_1 and accept a payoff of $-4/3$ since his opponent will receive a payoff of -4.

In Game 2 the threat potentials are just reversed; player Y can exercise a strong threat by convincing player X that he will always play Y_1. At best, player X can threaten to always play X_2.

The selection of the best threat strategy is not obvious and is often complicated by the fact that the best threat strategy may be a mixed strategy instead of a pure strategy. However, if payoffs are linearly transferable between the two players (i.e., if the transfer of one unit of payoff from player X results in a gain of one unit of payoff for player Y) the problem can be solved quite easily.

It can be shown that the optimum threat strategies for a nonzero-sum game are the same as the optimum strategies for the zero-sum game which is obtained by subtracting the payoffs for player Y from the payoffs for player X. Thus for Game 1 in Fig. 5-15, the optimum threat strategy is obtained by solving the following zero-sum game:

$$
X \begin{array}{c} \\ \end{array} \begin{bmatrix} X_1 \ (4-1) & (-4/3 + 4) \\ X_2 \ (-3+1) & (1-4) \end{bmatrix} = \begin{bmatrix} 3 & 8/3 \\ -2 & -3 \end{bmatrix}
$$

with columns Y_1, Y_2 under Y.

This game has a saddle point at $8/3$ thus the optimum threat solution is (X_1, Y_2).

The optimum threat strategy for Game 2 is obtained by solving the following zero-sum game:

$$
\begin{array}{cc} & Y \\ & Y_1 \ \ Y_2 \\ X & \begin{array}{cc} X_1 & -3 \ \ 8/3 \\ X_2 & -2 \ \ 3 \end{array} \end{array}
$$

Since -2 is a saddle point for this game, the optimum threat solution is (X_2, Y_1). It is reasonable to expect the bargaining advantage which a player holds when he has the strongest threat potential would be reflected in the solution of the game. Let us see if this is true for the games in Fig. 5-15. As illustrated in Fig. 5-16, both games have the same payoff polygon because Game 2 was obtained from Game 1 by simply switching the positions of two of the payoff pairs. As shown in Fig. 5-7, all games which have the same set of payoff pairs will have the same payoff polygon. The security levels for these games are obtained by finding the minimax solution for the appropriate zero-sum games which are shown in Fig. 5-16. For Game 1, the minimax strategy for player X is $(3/7, 4/7)$ and his security level is 0; player Y's minimax strategy is $(4/5, 1/5)$ and his security level is also 0. For Game 2, the minimax strategies are $(3/4, 1/4)$ for player X and $(1/2, 1/2)$ for player Y. The security level for both players is still 0.

If the security levels are used to determine the location of the status quo point, both Game 1 and Game 2 will have the same status quo point as well as the same payoff polygon. Therefore, regardless of the technique used to combine the status quo point and the payoff polygon to obtain a solution, both of the games in Fig. 5-15 will have the same solution. This solution ignores the fact that player X has the strongest threat potential in Game 1 and player Y has the same advantage in Game 2.

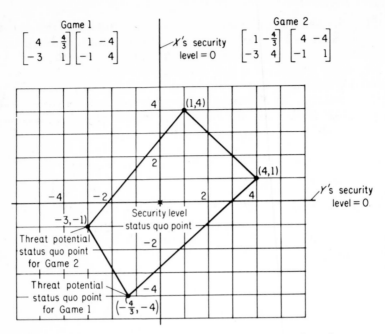

Fig. 5-16. Payoff polygon and status quo points for
Game 1 and Game 2.

To avoid this problem the pair of payoffs determined by the pair of threat strategies can be used as the status quo point. For Game 1 the players' best threat strategies were X_1 and Y_2 thus the threat potential status quo point for this game would be $(-4/3, -4)$. For Game 2 the best threat strategies are Y_1 and X_2 so the status quo point would be $(-3, -1)$. As we shall see in the following sections of this chapter, the use of these status quo points would lead to a more favorable solution for player X in Game 1 and a more favorable solution for player Y in Game 2.

Many different status quo points can be defined for a nonzero sum game. Such points can be obtained by applying different weights to the players' threat potentials and security levels or by introducing entirely different assumptions. For the sake of brevity, only the two status quo points which have been discussed above will be used in the following discussion of the techniques which are used to obtain negotiated solutions for nonzero sum games.

SOLUTION TECHNIQUES

Many techniques have been proposed for combining the status quo point and the negotiation set to obtain a solution for a negotiable game. The four techniques which we will examine in detail are: the global optimum, the 45° bisector, the maximization of the product of the payoffs, and the axiomatic solution of J. F. Nash.

Global Optimum. From a global point of view, the optimal negotiated solution is the pair of decisions which maximizes the total payoff to both players. The global optimum will be the nego-tiated solution for a nonzero sum game if the competitors decide to cooperate so as to obtain the largest total payoff for the conspiracy regardless of the payoffs to the individual players.

When the total payoff is the same at both endpoints of the negotiated solution, such as in the game illustrated in Fig. 5-16, the total payoff will be the same for all points in the negotiation set. In this situation the global-optimum technique provides no assistance in the selection of a particular solution point.

For a game, such as the one shown in Fig. 5-17, which has a different total payoff at each endpoint of the negotiation set, the

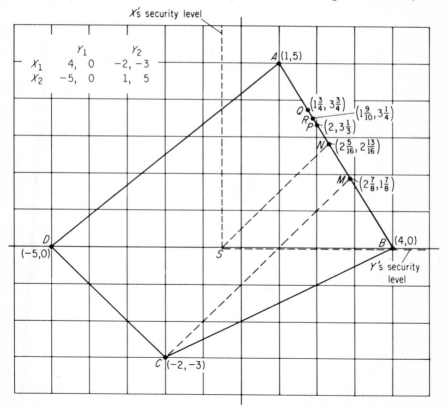

Fig. 5-17. Graphical solution of a nonzero-sum game.

global-optimum solution will occur at the endpoint of the negotiation set which has the largest total payoff. To induce his competitor to accept this solution, the player who receives the largest portion of the total payoff usually must offer to share a part of his payoff with his competitor. However, the global-optimum technique provides no guidelines for determining the amount of the transfer payment or side payment required to make a competitor agree to the global-optimum solution.

Since the global-optimum technique usually provides no guidelines for selecting an equitable solution for a negotiable game, the technique is of little interest from the point of view of game theory.

The 45° Bisector. It is possible to obtain a negotiated solution for a nonzero-sum game by a relatively simple graphical procedure. The steps in this procedure are

1. Draw the payoff polygon for the game.
2. Determine the negotiation set.
3. Select a status quo point.
4. Draw a 45° line from the status quo point to the line which represents the negotiation set.

The first three steps of the procedure have already been discussed in detail. The rationale for the last step is as follows. Assuming that the status quo point is the point from which the bargaining should begin, the 45° line adds equally to the payoffs of both players and is thus the most equitable solution.

The use of this procedure is illustrated in Fig. 5-17. The security level is 0 for player Y and $-\frac{1}{2}$ for X, thus the negotiation set is the line AB. If we use the security levels to determine the status quo point, we draw a 45° line from $(-\frac{1}{2}, 0)$ to the line AB; the two lines intersect at point N ($2\frac{5}{16}, 2\frac{13}{16}$). The slightly higher value of the game for Y can be attributed to his higher security level.

Player X would probably feel that this solution does not adequately reflect his superior threat potential. From X's point of view the pair of payoffs for the players' threat potential $(-2, -3)$ would be a superior status quo point. Using this point as the status quo point, the graphical procedure gives a negotiated solution of $(2\frac{7}{8}, 1\frac{7}{8})$, point M in Fig. 5-17.

It is frequently easier to determine the 45° bisector solution analytically than graphically. The steps in the procedure for computing the solution are

1. Find the equation for the negotiation line.
2. Determine the equation for the 45° line passing through the status quo point.

3. Find the point at which these lines intersect.

To illustrate this procedure we will determine the two solutions to the game shown in Fig. 5-17.

The equation for a line which passes through the points (X_1, Y_1) and (X_2, Y_2) can be determined from the two-point formula:[3]

$$\frac{Y - Y_1}{X - X_1} = \frac{Y_2 - Y_1}{X_2 - X_1} \tag{5-1}$$

For the game in Fig. 5-17, the negotiation set is a portion of the line AB which passes through the points $(1, 5)$ and $(4, 0)$. The equation for this line is obtained by simply substituting the appropriate values in Eq. (5-1).

If we let the coordinates of point A be (X_1, Y_1) and the coordinates of point B be (X_2, Y_2), the equation for line AB is computed as follows:

$$X_1 = 1, \quad Y_1 = 5$$
$$X_2 = 4, \quad Y_2 = 0$$

Therefore:

$$\frac{Y - 5}{X - 1} = \frac{0 - 5}{4 - 1}$$

$$3(Y - 5) = -5(X - 1)$$

$$3Y - 15 = -5X + 5$$

$$3Y + 5X = 20 \tag{5-2}$$

Before we can find the equation for the 45° line which passes through the status quo point we must define the *slope of a line*. The slope of a line is a measure of the amount by which the Y coordinate changes when the value of the X coordinate is increased by one unit. The slope is positive for a line which "leans to the right" because the value of the Y coordinate increases as the value of the X coordinate increases. For a line which "leans to the left" the slope is negative because the value of the Y coordinate decreases as the value of the X coordinate increases. The slope of a horizontal line is zero because there is no change in the Y coordinate as the X coordinate changes. The slope of a vertical line is not defined. Figure 5-18 illustrates some different values for the slope of a line.

In explaining the rational for the 45° bisector method we indicated that the 45° line was selected because movement along this

---

[3] A discussion of the formulas used for deriving the equation for a line can be found in any elementary geometry text.

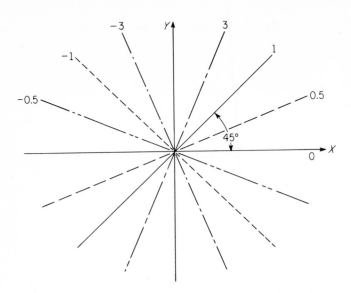

Fig. 5-18. Some examples of the slope of a line.

line adds equally to the payoff of both players; that is, the Y coordinate increases by one unit when the value of the X coordinate is increased by one unit. Thus the slope of the 45° line is equal to 1. Since we also know that the 45° line passes through the status quo point we can obtain the equation for the line from the point-slope formula.

$$\frac{Y - Y_0}{X - X_0} = \text{slope} \tag{5-3}$$

where (X_0, Y_0) are the coordinates of the status quo point.

For the game illustrated in Fig. 5-17 the security level status quo point is $(-\frac{1}{2}, 0)$ so the equation for the 45° line is computed as follows.

$$\frac{Y - 0}{X - (-\frac{1}{2})} = 1$$

$$Y = X + \frac{1}{2}$$

$$Y - X = \frac{1}{2} \tag{5-4}$$

The value of X and Y at the intersection point of these two lines must satisfy both Eq. 5-2 and Eq. 5-4. Thus we can determine the point of intersection by solving these equations simultaneously. To solve this set of equations we will multiply Eq. 5-4 by 5 and add it to Eq. 5-2.

$$5Y - 5X = 2\frac{1}{2} \qquad \text{Eq. 5-4 multiplied by 5}$$
$$3Y + 5X = 20 \qquad \text{Eq. 5-2}$$
$$8Y + 0 \ = 22\frac{1}{2} \qquad \text{Adding}$$
$$Y = {}^{45}\!/_{16} \qquad \text{Solving for } Y$$
$$= 2^{13}\!/_{16}$$

We can find the value of X from Eq. 5-4 as follows:

$$Y - X = \frac{1}{2} \qquad \text{Eq. 5-4}$$
$$X = Y - \frac{1}{2} \qquad \text{Solving for } X$$
$$X = 2^{13}\!/_{16} - \frac{1}{2}$$
$$= 2^{5}\!/_{16}$$

Therefore, when the security levels are used to determine the status-quo point, the 45° bisector solution is $(2^{5}\!/_{16}, 2^{13}\!/_{16})$ which is point N in Fig. 5-17.

The selection of a different status quo point does not change the equation for the negotiation line but does change the equation for the 45° line. Since the threat strategy status quo point has the coordinates $(-2, -3)$ the equation for the 45° line is computed as follows:

$$\frac{Y - Y_0}{X - X_0} = \text{slope} \qquad [5\text{-}3]$$
$$\frac{Y - (-3)}{X - (-2)} = 1$$
$$Y + 3 = X + 2$$
$$X - Y = 1 \qquad (5\text{-}5)$$

To find the point of intersection we solve Eq. 5-2 and 5-5 simultaneously.

$$3X - 3Y = \ 3 \qquad \text{Eq. 5-5 multiplied by 3}$$
$$5X + 3Y = 20 \qquad \text{Eq. 5-2}$$
$$8X + 0 \ = 23 \qquad \text{Adding}$$
$$X = {}^{23}\!/_8 \qquad \text{Solving for } X$$
$$= 2^{7}\!/_8$$

$$X - Y = 1 \qquad \text{Eq. 5-5}$$
$$Y = X - 1 \qquad \text{Solving for } Y$$
$$= 2\tfrac{7}{8} - 1$$
$$= 1\tfrac{7}{8}$$

The solution $(2\tfrac{7}{8}, 1\tfrac{7}{8})$ is the solution point M in Fig. 5-17.

Since we can define many different status quo points for a negotiated game there will be a large number of $45°$ bisector solutions for the game. Thus one of the major problems encountered in the use of this solution technique is the selection of the "best" status quo point. We shall examine this problem in more detail at the end of the chapter when we compare the results obtained from using different solution techniques.

Maximization of the Product of the Player's Payoffs. Let us assume that the players of a nonzero-sum game have proposed two solutions, (X_1, Y_1) and (X_2, Y_2) such that $X_1 > X_2$ and $Y_1 < Y_2$. Thus player X will prefer the solution (X_1, Y_1) and player Y will prefer (X_2, Y_2). We now pose a question: Would it be equitable to move from solution (X_1, Y_1) to solution (X_2, Y_2)? It is reasonable to expect that such a move would be an equitable bargain if the relative loss of payoff for player X as a result of the change from (X_1, Y_1) to (X_2, Y_2) would be less than the relative loss of payoff for player Y which would result in moving from the (X_2, Y_2) to (X_1, Y_1).

Player X's relative loss can be measured by dividing his loss of payoff by his payoff in his preferred position.

$$\text{Relative loss for player } X = \frac{(X_1 - X_2)}{X_1}$$

Player Y's relative loss is computed in a similar way:

$$\text{Relative loss for player } Y = \frac{(Y_2 - Y_1)}{Y_2}$$

Thus the condition for moving from solution (X_1, Y_1) to (X_2, Y_2) is

$$\frac{X_1 - X_2}{X_1} \leqslant \frac{Y_2 - Y_1}{Y_2}$$

or

$$X_1 Y_2 - X_2 Y_2 \leqslant X_1 Y_2 - X_1 Y_1$$

This equation is valid only when $X_2 Y_2 \geqslant X_1 Y_1$.

The only point on the line AB in Fig. 5-17 for which this condition is always satisfied is the point at which the product of the players' payoffs is a maximum. We can find the coordinates of this point either graphically or analytically.

To find this point graphically we first solve the equation for line AB (Eq. 5-2) for Y.

$$3Y + 5X = 20 \qquad\qquad [5\text{-}2]$$

$$3Y = 20 - 5X$$

$$Y = 20/3 - {}^5\!/_3 X \qquad\qquad (5\text{-}6)$$

By substituting a selected value of X in Eq. 5-6 we obtain the corresponding value for Y. We then multiply the value of X by the corresponding value for Y to obtain the desired product. The process of locating the value of X which will maximize the value of the product of X and Y is greatly expedited by plotting a graph such as Fig. 5-19. As shown in Fig. 5-19, the product of the players' payoffs is a maximum when X is equal to 2; the corresponding value of Y is $3^1/_3$. Thus, the maximum product solution for our example is the point $(2, 3^1/_3)$ which is shown as point P in Fig. 5-17.

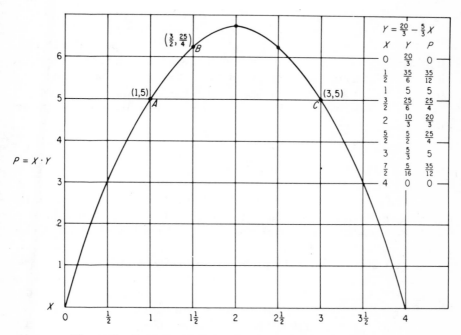

Fig. 5-19. Graph of the product of the players' payoffs.

To analytically determine the value of X which maximizes the product of the players' payoffs we must use the techniques of differential calculus. First we must discuss the *slope of a curved graph*. As shown in Fig. 5-18, the slope of a straight line is the same at every point on the line. But a curved graph, such as the graph in Fig. 5-19, does not have the same slope at every point. We can compute the slope of the line passing through any two points on the graph from the following formula:

$$\text{Slope} = \frac{Y_2 - Y_1}{X_2 - X_1} = \frac{\Delta Y}{\Delta X} \qquad (5\text{-}7)$$

For example, the slope of the line passing through the points $(1, 5)$ and $(1\frac{1}{2}, 6\frac{1}{4})$ in Fig. 5-19 is

$$\text{Slope} = \frac{6\frac{1}{4} - 5}{1\frac{1}{2} - 1} = \frac{5/4}{1/2} = \frac{5}{2} = 2\frac{1}{2}$$

Since the slope is changing at every point, Eq. 5-7 can yield only the average slope between the two points. If we want to compute the slope at a single point we can *approach* the correct answer by computing the slope of the line which passes through two points which are very close. We cannot compute the true value of the slope at a single point with this formula because when there is only a single point, $\Delta X = 0$, so it is not possible to evaluate the ratio $\Delta Y/\Delta X$.

The slope of a curve at a single point is the slope of the straight line which is tangent to the curve at the point. The tangent at a point is the line that just touches the curve at the point. The slope of the tangent line at a point is the limit which the ratio $\Delta Y/\Delta X$ *approaches* as ΔX becomes very small and *approaches* zero. We use the symbol dY/dX to represent this limit.

$$\frac{dY}{dX} = \lim_{\Delta X \to 0} \frac{\Delta Y}{\Delta X}$$

We call dY/dX the *derivative of Y with respect to X*. The value of the derivative at a point is equal to the slope of the curve at that point.

The slope of the tangent line has a positive value if the line leans to the right [point $(1, 5)$ in Fig. 5-19] and a negative value if the tangent leans to the left [point $(3, 5)$ in Fig. 5-19]. It can be seen from Fig. 5-19 that a maximum will occur whenever the slope of the tangent line changes from a positive to negative value. Furthermore, the tangent line at the maximum point will be a horizontal line which has a slope of zero. Therefore, if we have an equation for the derivative of a function (such as the function in Fig. 5-19) we can find the maximum point for the function by setting the derivative to zero and solving the resulting equation for the desired value of X.

The basic rule for finding the derivative of a simple polynomial function can be stated as follows:

$$\text{If} \quad Y = aX^n$$

$$\frac{dY}{dX} = n \cdot aX^{n-1}$$

For example,

$$\text{If} \quad Y = 3X^2$$

$$\frac{dY}{dX} = 2 \cdot 3X^{2-1}$$

$$\frac{dY}{dX} = 6X$$

The preceding discussion illustrates that the coordinates of a maximum point can be determined by the following procedure.

1. Obtain the algebraic expression for the function for which you want to obtain the maximum.
2. Determine the derivative of the function.
3. Set the derivative of the function equal to zero and solve the resulting equation for the desired value of X.
4. Compute the value of Y corresponding to the value of X which was obtained in Step 3.

This procedure is used below to obtain the coordinates for the maximum point of the graph in Fig. 5-19.

1. The expression for the product of the players' payoffs, P, is obtained as follows:

$$P = XY$$

Since

$$Y = (20/3 - 5/3X) \tag{5-6}$$

$$P = X(20/3 - 5/3X)$$

$$= 20/3X - 5/3X^2 \tag{5-8}$$

2. Obtain the derivative of Eq. 5-8 with respect to X:

$$P = 20/3X - 5/3X^2$$

$$\frac{dP}{dX} = 1 \cdot 20/3X^{1-1} - 2 \cdot 5/3X^{2-1}$$

$$= 20/3 - 10/3X \tag{5-9}$$

3. Set the derivative (Eq. 5-9) equal to zero and solve the resulting equation:

$$0 = 20/3 - 10/3X$$
$$10/3X = 20/3$$
$$X = 20/3 \cdot 3/10$$
$$= 2$$

4. Compute the corresponding value of Y from Eq. 5-6.

$$Y = 20/3 - 5/3X \qquad\qquad [5\text{-}6]$$
$$Y = 20/3 - 5/3 \cdot 2$$
$$= 10/3$$

The value of the product of the player's payoffs, P, can be obtained by simply multiplying the value of X by the corresponding value of Y or by substituting the value of X in Eq. 5-8.

$$P = 20/3X - 5/3X^2 \qquad\qquad [5\text{-}8]$$
$$= 20/3 \cdot 2 - 5/3 \cdot 4$$
$$= 20/3$$

or

$$P = X \cdot Y$$
$$= 2 \cdot 10/3 = 20/3$$

The solution $(2, 3\frac{1}{3})$ is marked as point P in Fig. 5-17. It should be noted that if the solution is moved from point P to any other point on line AB, the *relative* loss or gain in the payoff for player X will be equal to the *relative* gain or loss in the payoff for player Y. For example, in moving from point P to point A in Fig. 5-17, player X loses $\frac{1}{2}$ of his current payoff of 2 and player Y gains $\frac{1}{2}$ of his current payoff of $3\frac{1}{3}$. Similarly, in moving from point P to point M, the gain in the payoff for player X is equal to $\frac{7}{16}$ of his payoff at point P and the loss for player Y is equal to $\frac{7}{16}$ of his payoff at point P. The equality of relative gains or losses in payoffs is one of the major advantages which is cited by the proponents of the maximum product technique.

The Axiomatic Solution of J. F. Nash. In the axiomatic approach to the solution of negotiated games, the first step is to specify the characteristics that are desired in the solution. Then a general method is sought for obtaining a solution which has the desired characteristics. In 1950 J. F. Nash proposed four general conditions which he felt should be satisfied by any "acceptable" solution for a nonzero-sum game.[4] Each of these conditions is discussed below. Techniques which can be used to obtain solutions which satisfy all four conditions, are then examined.

1. *The solution should not depend on the way the players are labeled.* The ability to negotiate successfully may be dependent on exogenous variables such as differences in the stock of capital which is available to the players or differences in the negotiating ability of the players. Nash's first condition eliminates such factors from the analysis. We must be concerned solely with the differences in bargaining power that result from the characteristics of the payoff matrices for the players.

For example, consider the two games shown in Fig. 5-20.

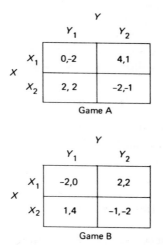

Game A

Game B

Fig. 5-20. Games in which the roles of the players are interchanged.

These games are exactly alike except that the roles of the two players are interchanged. In Game A if player X selects X_1 the payoffs to player Y will be - 2 if he chooses Y_1 and 1 if he chooses Y_2. In Game B if player Y chooses Y_1 the payoff to player X will be - 2 if he selects X_1 and 1 if he selects X_2. Nash's first condition specifies that the solutions for these games should be the same except that the payoffs for the players should be interchanged.

2. *The solution should be in the negotiation set.* This condition guarantees that the payoff to either player will not be less than his

[4] J. F. Nash, "The Bargaining Problem," *Econometrica*, Vol. 18 (1950), pp. 155–162.

security level and that both players will not be able to simultaneously improve their payoffs by moving from the solution point.

3. *The solution should not be affected by a linear transformation of the payoffs.* The general form for a linear transformation is

$$Y^* = a + bY \qquad (5\text{-}10)$$

Linear transformations are frequently used to transfer from one unit of measurement to another. For example, to convert temperature measurements from degrees Fahrenheit to degrees Centigrade the linear transformation is

$$T_c = -\frac{160}{9} + \frac{5}{9}T_F$$

This condition does *not* mean that the solution point will not change if the payoffs are transformed. It does mean that the relative payoffs will remain the same. The two games which are diagrammed in Fig. 5-21 are exactly the same except that the payoffs for Game B were obtained by the following linear transformation of the payoffs for Game A.

$$P_B = 2P_A + 3$$

The third criterion specifies that if the solution to Game A is at point C, the solution to game B, which is at C^*, must be such that

$$\frac{A^*C^*}{C^*B^*} = \frac{AC}{CB}$$

As shown in Fig. 5-21, these ratios have a value of $5/3$ for the solutions at point C and point C^*.

Condition 3 can also be stated in terms of the players' strategies instead of their payoffs. To achieve the desired payoffs $(4^1/2, 3^1/2)$ for Game A in Fig. 5-21 the players should make their choices so that the combination (X_1, Y_2) occurs part of the time and (X_2, Y_1) occurs the rest of the time. Let p be the percent of the time that combination (X_1, Y_2) is chosen. Then

$$\text{Player } X\text{'s payoff} = 4^1/2 = 6p + 2(1-p)$$

$$4p = 2^1/2$$

$$p = 5/8 \quad \begin{array}{l}\text{[percent of time} \\ (X_1, Y_2) \text{ is chosen]}\end{array}$$

$$1 - p = 3/8 \quad \begin{array}{l}\text{[percent of time} \\ (X_2, Y_1) \text{ is chosen]}\end{array}$$

A similar calculation will show that the same probabilities should be used to achieve the desired payoffs for Game B. Thus the third

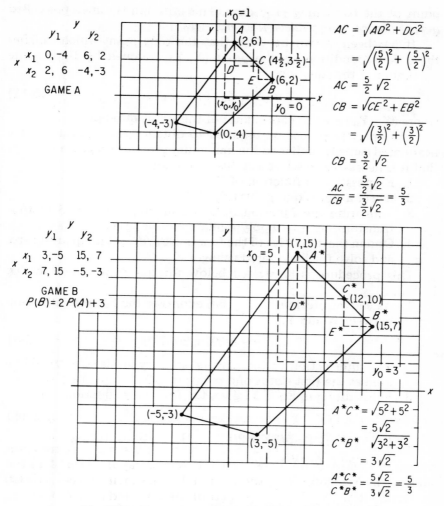

Fig. 5-21. Two games with linearly related payoffs.

condition requires that the strategies which the players use to achieve the desired payoffs should not be affected by a linear transformation of the payoffs.

4. *If additional payoff pairs become available, but the status quo point remains unchanged, the additional payoff pairs should either be included in the solution of the game or else they should not affect the solution of the game.* This criterion implies that an alternative which is rejected by both sides in a negotiated game should not affect the solution of the game unless the alternative changes the reference

point of the bargaining procedure. This criterion is sometimes called *independence from irrelevant alternatives.*

It has been shown by Nash that the only point which satisfies these four conditions is the point (X, Y) obtained by finding the maximum of the function

$$Z = (X - X_0)(Y - Y_0) \tag{5-11}$$

where (X_0, Y_0) are the coordinates of the status quo point.

The analytic procedure required to obtain Nash's solution for a negotiated game is similar to the maximum-product procedure except that it is necessary to select a status quo point.

1. Express Y as a function of X.
2. Select a status quo point (X_0, Y_0).
3. Substitute the function obtained in Step 2 in Eq. 5-11 thus transforming it to a polynomial function in X.
4. Obtain the derivative of the function dZ/dX set it equal to zero, and solve for the desired value of X.

This procedure is illustrated below for the game shown in Fig. 5-17.

1. As shown previously, Eq. 5-6 expresses the linear relationship between X and Y.

$$Y = -5/3X + 20/3 \qquad [5\text{-}6]$$

2. The intersection of the players' security levels $(-\frac{1}{2}, 0)$ will be used as the first status quo point; thus $X_0 = -\frac{1}{2}, Y_0 = 0$.
3. Substituting in Eq. 5-11 gives the following equation:

$$Z = (X - X_0)(Y - Y_0) \qquad [5\text{-}11]$$
$$= [X - (-\frac{1}{2})][(-5/3X + 20/3) - 0] \qquad \text{Substituting 5-6}$$
$$= (X + \frac{1}{2})(-5/3X + 20/3)$$
$$Z = -5/3X^2 + 35/6X + 10/3 \qquad (5\text{-}12)$$

4. The derivative of Eq. 5-12 is

$$\frac{dZ}{dX} = -10/3X + 35/6$$

If we let $\dfrac{dZ}{dX} = 0$, then

$$0 = -10/3X + 35/6$$
$$10/3X = 35/6$$
$$X = 35/6 \times 3/10$$
$$= 1^3/_4$$

5. The corresponding value of Y is found from Eq. 5-6:

$$Y = -5/3X + 20/3$$
$$= -5/3 \times 7/4 + 20/3$$
$$= -35/12 + 80/12$$
$$= 45/12$$
$$= 3^3/4$$

This solution is shown as point Q in Fig. 5-17. The same technique can be used to obtain a solution for which the threat potentials $(-2, -3)$ are used as the status quo point. In this case,

$$(X_0, Y_0) = (-2, -3)$$
$$Z = -5/3X^2 + 19/3X + 58/3$$
$$\frac{dZ}{dX} = -10/3X + 19/3$$
$$X = 1^9/_{10}$$
$$Y = 3^1/_2$$

The threat potential solution is shown as point R in Fig. 5-17.

Nash's solution can also be obtained by the following graphical procedure.

1. Determine the two payoff points through which the negotiation line passes.
2. Using the two points determined in step 1, find the linear transformations for the X and Y coordinates which will transform the payoff pair preferred by player X to the new pair of payoffs $(1, 0)$ and the payoff pair preferred by player Y to the new pair of payoffs $(0, 1)$. Apply the same pair of linear transformations to all other entries in the payoff matrix.
3. Draw the payoff polygon and select a status quo point.
4. Draw a 45° line from the status quo point to the negotiation line and determine the coordinates of the point of intersection.
5. Translate the coordinates of the intersection point back into the original coordinate system using the linear transformations determined in step 2.

This procedure is illustrated below for the game in Fig. 5-17.

1. The negotiation line passes through the points A $(1, 5)$ and B $(4, 0)$.

2. Since player X prefers point B and player Y prefers point A, we must find a linear transformation for the X and Y coordinates which will change the payoffs at point A from $(1, 5)$ to $(0, 1)$ and the

payoffs at point B from $(4, 0)$ to $(1, 0)$. The general form for a linear transformation is

$$Y^* = a + bY \qquad\qquad [5\text{-}10]$$

where Y is the present payoff and Y^* the desired payoff.

Using the present and desired values for X and Y we can write a set of equations which we can solve simultaneously to determine the values for the coefficients in the transformation equations.

At point A the payoff for player Y must be changed from 5 to 1. Therefore

$$1 = a + 5b \qquad\qquad (5\text{-}13)$$

At point B the payoff for player Y is already 0 and we want it to be 0 after the transformation, so

$$0 = a + 0b \qquad\qquad (5\text{-}14)$$

Solving Eq. 5-14 for a we find that

$$a = 0$$

Substituting this value of a in Eq. 5-13 and solving for b gives

$$1 = 0 + 5b$$

$$b = \text{\textonesuperior/s}$$

Therefore, the desired linear transformation for player Y's payoffs is

$$Y^* = \text{\textonesuperior/s}\, Y \qquad\qquad (5\text{-}15)$$

For player X we obtain the following equations:

At point A, $\qquad\qquad 0 = a + b \qquad\qquad (5\text{-}16)$

At point B, $\qquad\qquad 1 = a + 4b \qquad\qquad (5\text{-}17)$

Solving these equations,

$$a = -\text{\textonesuperior/3}$$

$$b = \text{\textonesuperior/3}$$

The desired linear transformation for player X's payoffs is

$$X^* = -1/3 + 1/3X \qquad\qquad (5\text{-}18)$$

By applying these transformations to all the pairs of payoffs in the game in Fig. 5-17 we obtain the following transformed game:

	Y_1	Y_2
X_1	1, 0	- 1, - 3/5
X_2	- 2, 0	0, 1

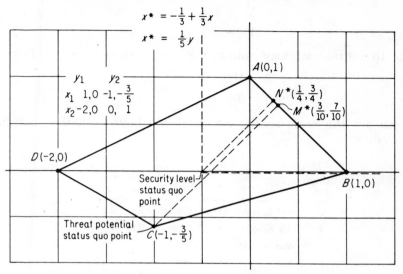

The game was obtained by applying
the following linear transformation to the game in Fig. 5-17

$$x^* = -\tfrac{1}{3} + \tfrac{1}{3}x$$

$$x^* = \tfrac{1}{5}y$$

	y_1	y_2
x_1	1,0	$-1,-\tfrac{3}{5}$
x_2	$-2,0$	0, 1

$A(0,1)$

$N^*(\tfrac{1}{4},\tfrac{3}{4})$

$M^*(\tfrac{3}{10},\tfrac{7}{10})$

$D(-2,0)$

Security level—
status quo
point

$B(1,0)$

Threat potential
status quo point $C(-1,-\tfrac{3}{5})$

Fig. 5-22. Graphical method for obtaining Nash's solution.

3. The payoff polygon for the transformed game is drawn in Fig. 5-22. The player's security levels have not been changed ($X_0 = -\tfrac{1}{2}$, $Y_0 = 0$) but the payoffs for the threat strategy are now ($-1, -\tfrac{3}{5}$). We will find the solution using both of these status quo points.

4. The 45° bisector lines are shown in Fig. 5-22. The intersection points, which can be determined either graphically or analytically, are ($\tfrac{1}{4}$, $\tfrac{3}{4}$) if the security levels are used to determine the status quo point and ($\tfrac{3}{10}$, $\tfrac{7}{10}$) if the threat potential status quo point is used.

5. The coordinates for the intersection points are transformed back into the original coordinate system using Eq. 5-15 for player Y's payoffs and Eq. 5-18 for player X's payoffs.

$$X^* = -\tfrac{1}{3} + \tfrac{1}{3}X \qquad [5\text{-}18]$$

Therefore

$$X = 3X^* + 1$$

$$Y^* = \tfrac{1}{5}Y \qquad [5\text{-}15]$$

Therefore

$$Y = 5Y^*$$

For the security level solution, $X^* = {}^1/_4$ and $Y^* = {}^3/_4$. Thus

$$X = 1{}^3/_4$$
$$Y = 3{}^3/_4$$

For the threat strategy solution $X^* = {}^3/_{10}$ and $Y^* = {}^7/_{10}$. Thus

$$X = 1{}^9/_{10}$$
$$Y = 3{}^1/_2$$

These are the same results we obtained above with the analytical procedure.

Solution Point	Technique	Status Quo Point	Payoffs x	y
M	45° bisector	Threat potentials	$2\frac{7}{8}$	$1\frac{7}{8}$
N	45° bisector	Security levels	$2\frac{5}{16}$	$2\frac{13}{16}$
P	Max. product	---	2	$3\frac{1}{3}$
Q	Nash	Security levels	$1\frac{3}{4}$	$3\frac{3}{4}$
R	Nash	Threat potentials	$1\frac{9}{10}$	$3\frac{1}{2}$

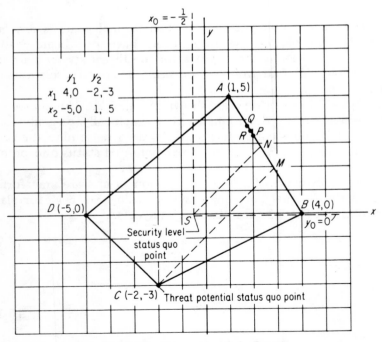

Fig. 5-23. Game A—standard game.

SUMMARY

The solutions for three games are illustrated in Figs. 5-23, 5-24, and 5-25 to summarize the techniques which have been discussed in this chapter and to indicate some of the difficulties which are encountered in developing solution techniques for negotiated games.

Game A (Fig. 5-23) is the same as the game in Fig. 5-17. This game will be used as a "standard." The other games are slight deviations from Game A. To form Game B (Fig. 5-24) the payoffs for the decision pair (X_1, Y_2) were changed from $(-2, -3)$ to $(-2, -5)$. To

Solution Point	Technique	Status Quo Point	Payoffs x	y
M	45° bisector	Threat potentials	$3\frac{5}{8}$	$\frac{5}{8}$
N	45° bisector	Security levels	$2\frac{5}{16}$	$2\frac{13}{16}$
P	Max. product	---	2	$3\frac{1}{3}$
Q	Nash	Security levels	$1\frac{3}{4}$	$3\frac{3}{4}$
R	Nash	Threat potentials	$2\frac{1}{2}$	$2\frac{1}{2}$

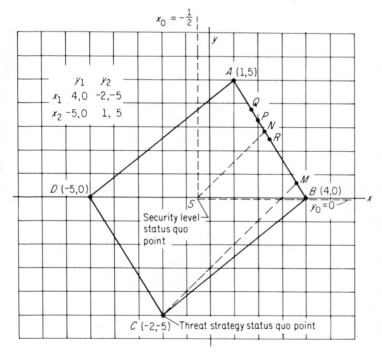

Fig. 5-24. Game B—game with a greater threat potential for player X.

Solution Point	Technique	Status Quo Point	Payoffs x	y
M	45° bisector	Threat potentials	$2\frac{7}{8}$	$1\frac{7}{8}$
N	45° bisector	Security levels	$2\frac{1}{2}$	$2\frac{1}{2}$
P	Max. product	---	2	$3\frac{1}{3}$
Q	Nash	Security levels	2	$3\frac{1}{3}$
R	Nash	Threat potentials	$1\frac{9}{10}$	$3\frac{1}{2}$

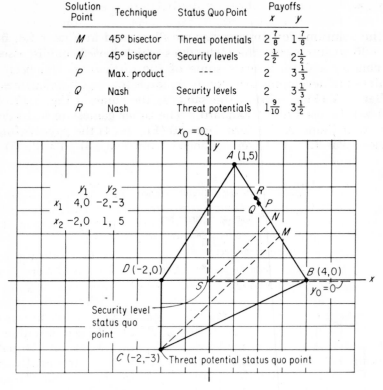

Fig. 5-25. Game C—game with higher security level for player Y.

form Game C (Fig. 5-25) the payoffs for decision pair (X_2, Y_1) were changed from $(-5, 0)$ to $(-2, 0)$.

These three games illustrate that none of the solutions which we have obtained are completely satisfactory because it is possible to make changes in the payoff matrix which result in changes in the bargaining strength of the players but which are not reflected in the solution for the game.

For example, changing the decision pair (X_1, Y_2) from $(-2, -3)$ in Game A to $(-2, -5)$ in Game B increased player X's threat potential. In both games player X would threaten to play X_1. In Game A player Y might threaten to play Y_2 because his loss of 3 is not much greater than X's loss to 2. But in Game B player Y cannot select Y_2 without risking a loss of 5 which is $2\frac{1}{2}$ times as large as the loss that his competitor would suffer.

As can be seen by comparing Figs. 5-23 and 5-24, only two of the five solutions reflect player X's increased threat potential. In both of these solutions the threat potentials are used to determine the location of the status quo point. The other three solutions are not changed because the change in the payoff matrix does not change the players' security levels or the location of the negotiation line.

Similar results are obtained for Game C when the change in the payoff matrix increased the security level for player X but did not change the security level for player Y and did not change the threat potential for either player. In this case only the two solutions which use the security levels to determine the location of the status quo point were affected. As can be seen by comparing Figs. 5-23 and 5-25 for the two solutions, the payoff for player X increased as would be expected.

Whenever the solution technique requires the use of a status quo point the analyst must select the most relevant status quo point. Game theory does not provide any assistance in the selection of the "best" status quo point.

One of the advantages of the maximum-product solution technique is the fact that is does not require the selection of a status quo point. In fact, the maximum-product solution depends only on the equation for the negotiation line. Since the negotiation line was the same for Games A, B, and C the maximum-product solution was also the same, even though there were significant changes in either the threat potentials or the security levels. The failure to adequately reflect the players' advantages and disadvantages is frequently cited as an important weakness of the maximum-product solution technique.

It should be clear that the different solutions for a negotiated game occur because different theories of bargaining and arbitration have been employed in attempting to solve the problem. The examples given above help demonstrate the flexibility of the ideas involved in the concepts of bargaining and arbitration as well as the difficulties involved in utilizing these concepts to formulate rational solutions for conflict situations when negotiations are allowed.

CHAPTER 6

n-Person Games

All of the games we have studied so far have involved only two opposing interests. In this chapter we will analyze games which have more than two players. These games are called *n-person games.*

Two major assumptions will be made throughout this chapter. First it will be assumed that all of the players in a game are able to communicate and negotiate with each other and, if they desire, to cooperate with each other. If a group of players decide to cooperate we say they form a *coalition.* As we will see, the analysis of coalition formation is of prime importance in the study of *n*-person games. The second major assumption which will be made is that the players are able to make *side payments.* A side payment is simply a transfer of payoffs between players.

Since the players can make side payments, they will form a coalition if the payoffs are such that the members of the coalition can, through cooperation, achieve a higher total payoff for the coalition than they can by playing as individuals. After the coalition has maximized its total payoff, the payoffs to the members of the coalition can be adjusted by making side payments.

PAYOFF MATRIX FOR n-PERSON GAMES

If we assume that the payoffs for the players of a two person nonzero-sum game are paid by the "house," we can treat the game as a three-person zero-sum game. The third player, the house, makes no strategic choices so we can represent this three-person game by a two-dimensional payoff matrix in which the payoff for each pair of choices is an ordered triple. The first entry in the triple is the payoff for player X, the second entry the payoff for player Y, and the third entry the payoff for the house. The payoff matrix for a three-person zero-sum game is shown in Fig. 6-1.

The procedure described above can be used to change any *n*-person nonzero-sum game into an $n + 1$ person zero-sum game. The

Y

	Y_1	Y_2	Y_3
X X_1	(5,0,-5)	(5,-2,-3)	(3,-5,2)
X_2	(0,5,-5)	(2,2,-4)	(-1,7,-6)

Fig. 6-1. Payoffs for a three-
person zero-sum game.

representation of such an n-person game would be an n-dimensional matrix which has an $n + 1$ dimensional payoff vector for each combination of the players' choices. Since the payoff for the house is always the negative of the sum of the payoffs of all the players, it is not necessary to include the last entry in each payoff vector. However, to avoid confusion, we will always include the entry for the house in the vector of payoffs.

A large part of n-person game theory has been developed for *constant-sum games*. A constant-sum game is simply a game for which the sum of the entries in each payoff vector is equal to the same constant. For example, in the constant-sum game in Fig. 6-2, the sum of the payoffs for each pair of choices is 30.

A zero-sum game is, of course, a special type of constant sum game. Any n-person zero-sum game can be changed to a constant sum game simply by adding the same fixed amount to each player's payoffs. For example, the zero-sum game illustrated in Fig. 6-1 can be changed to the constant-sum game illustrated in Fig. 6-2 by adding 10

Y

	Y_1	Y_2	Y_3
X X_1	(15,10,5)	(15,8,7)	(13,5,12)
X_2	(10,15,5)	(12,12,6)	(9,17,4)

Fig. 6-2. Payoff matrix for a
three-person constant-sum game.

to the payoff for each player. Similarly, an n-person game with a constant sum, say K, can be converted to a zero-sum game by subtracting K/n from the payoffs for each player or by adding one more player with a payoff of $-K$. Thus the game in Fig. 6-2 can be changed to a zero-sum game by subtracting 10 from the payoffs for each player or by adding a new player whose payoff is always - 30.

It should be noted that the constant sum game in Fig. 6-2 could be changed to another zero-sum game by subtracting 15 from the payoffs for player X, 10 from the payoffs for player Y, and 5 from the payoffs for the house. It should be clear that it is possible to form many different zero-sum games from any constant-sum game by this simple process of subtraction. In fact, if fractional payoffs are permitted, an infinite number of different zero-sum games can be generated in this way.

COALITIONS

The conversion of a two-person nonzero sum game to a zero-sum game by the addition of a third player eliminates some of the difficulties involved in the analysis of nonzero-sum games. However, the existence of more than two players introduces a new type of difficulty, namely the problem of coalitions. A *coalition* is an agreement among some of the players to coordinate their available strategies in such a way that all of the members of the coalition will benefit. For example, for the game in Fig. 6-1, the players could decide to always adopt the strategy (X_2, Y_2) thus guaranteeing each player a positive payoff for each play of the game. The players could achieve an even greater payoff if they agreed to alternate between strategies (X_1, Y_1) and (X_2, Y_1) thus achieving an average payoff of 2½ for each player on each play of the game. If they use side payments the players can earn an even higher payoff by always selecting strategy (X_2, Y_3) thus providing an average payoff of 6 for the coalition and an average payoff of 3 for each player if they agree to divide the total payoff equally.

In a zero-sum game with two active players and the house only one coalition is possible. Therefore the analysis of such a game seeks to answer two questions: (1) Will there be a coalition? and (2) What strategy should the coalition adopt?

The addition of more players quickly complicates the analysis of coalition formation. If there are three active players, it is possible to form four different coalitions (XY, XZ, YZ, XYZ). If we assume that the players will form into two opposing groups, the players in a four-person game can form eight different coalitions as shown in Fig. 6-3. In general, there are 2^{n-1} possible ways of splitting N players into two groups.

Group 1	Group 2
WXYZ	No one
WXY	*Z*
WXZ	*Y*
WYZ	*X*
XYZ	*W*
WX	*YZ*
WY	*XZ*
WZ	*XY*

Fig. 6-3. The eight feasible coalitions for a game with four decision makers.

The above discussion indicates that one of the primary concerns of *n*-person game theory is the question of coalition formation and its effect on the outcome of the game.

COALITION FORMATION

Let us assume that a game has N players. Without loss of generality we can assume that S players form a coalition, which we will call the INS, and that the remaining $(N - S)$ players form a counter-coalition, which we will call the OUTS. The formation of the INS coalition is in accordance with common sense providing the game's payoffs are such that the total payoff for the coalition is greater than the sum of the payoffs for the S players if they played by themselves. Although it is not necessary for the remaining players to form the OUTS coalition, they have nothing to lose by such an action. The members of the OUTS can always play their strategies the same way as they would have played them independently so they can do at least as well collectively as they can individually. Moreover, they *may* be able to coordinate their strategies to their collective advantage.

Once the INS have formed, the OUTS will form in response to the formation of the INS. However, during the formation of the INS the players who are not included in the proposed coalition may try to prevent its formation by offering to share part of their payoffs with a potential member of the INS, thus hoping to induce the player to join a new coalition. Thus the ability and willingness of the players to make side payments will greatly affect the formation of coalitions. In fact, if we make the usual game theory assumption that the personal relations among the players do not influence their decisions, the membership of the coalitions will be determined entirely by the nature of the payoff matrix and the willingness of the players to make side payments.

Before examining coalition formation in more detail let us consider two examples which illustrate some of the problems which are encountered in the formation of a coalition.

Example 1. "Two for the Money." Players A, B, and C are asked to form a coalition. If any two of them are successful in forming a coalition the third player must pay each of them $1. If no coalition is formed, or if a three-player coalition is formed, there is no payoff at all.

We can analyze this game by listing the value of the game for all possible coalitions. Such a listing, which is called the characteristic function of an *n*-person game, is shown in Fig. 6-4.

Coalition	Value
A	0
B	0
C	0
AB	2
AC	2
BC	2
ABC	0

Fig. 6-4. The characteristic function
for "Two for the Money."

Although there is no way to select the "best" of the three two-player coalitions, we can reasonably assume that one of the two-man coalitions will be formed. The payoffs associated with these three two-man coalitions may be considered to be a "solution" of the game. We will examine such solutions in more detail later.

Now let us change the game slightly so that if coalition AB is formed, player C must pay player A $1.50 and player B $.50. Although this change does not increase the total payment to any coalition, player A seems to have an advantage. If coalition AB is formed, player A will earn more than he did in the previous game; if any other coalition is formed his payoffs will be the same as in the previous game. However, it is now almost impossible for player A to form coalition AB because both player B and player C will be better off if they form coalition BC. By forming coalition BC, player B will increase his payoff from $.50 to $1.00 and player C will increase his payoff from 0 to $1.00. Thus player A's greater payoff is really a disadvantage because it induces both of his competitors to exclude him from the coalition. Player A can remedy the situation by offering to make a side payment of $.50 to player B if coalition AB is formed. The game is then reduced to the previous game.

Example 2. "Divide the Pot." Consider the problems faced by three men who are given $30 which they can keep provided they can decide how to divide it by majority vote. The characteristic function for this game is shown in Fig. 6-5.

Coalition	Value
A	0
B	0
C	0
AB	30
AC	30
BC	30
ABC	30

Fig. 6-5. The characteristic function
for "Divide the Pot."

If each man voted to keep the entire amount for himself, no one would have a majority and the value of the game would be 0. Thus we can assume that some type of coalition will be formed. It makes little sense for a three-man coalition to form because the three players together can earn no more than two players but they will have to split the total payoff three ways. Although we cannot decide which pair of players will form a coalition it is reasonable to expect that some two-man coalition will be formed.

Let us assume that players A and B tentatively decide to form a coalition but that the negotiations are still open. Player C, assuming that A and B have settled on a 50–50 split, may now propose to B that they form coalition BC with payoffs (0, 18, 12). Player B may be tempted to hold out for a more favorable payoff such as (0, 20, 10) but, realizing that player C could form a coalition with A, player B decides to accept the offer made by player C before he changes his mind.

Now player A, realizing that he has been outsmarted, offers to give player B 70 percent of the total payoff if he will again join him in forming coalition AB. Since player B is interested solely in his payoff, he accepts player A's offer and the payoffs become (9, 21, 0). Player B is ready to end negotiations, but player C now approaches A and proposes the formation of coalition AC with a payoff of (12, 0, 18). Since both A and C will benefit from this proposal, the new coalition AC is formed and B is once again frozen out. But player B may now approach A and propose the formation of the original coalition AB with an equal split of the total payoff. Both A and B will gain from this proposal so the coalition AB with payoffs (15, 15, 0) is formed and we are back where we started.

It should be clear from the above example that efforts to obtain a single solution for an n-person game result in the same sort of circular reasoning that caused trouble in the zero-sum game without a saddle point. In the case of the zero-sum game the concept of the mixed strategy provides a solution to this difficulty. So far, no such simple solution has been discovered for the problem of shifting coalitions. However, n-person game theory does provide techniques for selecting a "set of acceptable solutions." In the remainder of this chapter we will examine some of these solution techniques.

SOLUTION TECHNIQUES FOR THE n-PERSON GAME

Let us assume that the game is a constant-sum game with N players. As indicated previously, we will also assume that the players divide into two groups with S players in the INS coalition and $(N - S)$

players in the OUTS coalition. After these coalitions have been formed, the game can be treated as a two-person game between the INS and the OUTS. The value of the game, which is listed in the characteristic function for the game, is the maximin value for the INS coalition—that is, the minimum total payoff that the members of the INS coalition can obtain regardless of the action taken by the members of the OUTS coalition. Since the game is a constant-sum game,

Value for the INS + value for the OUTS = Constant

Imputations. An imputation for an *n*-person game is any payoff vector for the game which satisfies two criteria:
1. The sum of the individual payoffs must equal the total payoff for the game.
2. The payoff for each player must be greater than or equal to the payoff which the player can attain for himself.

For example, the payoff vector (0, 9, 21) is an imputation for the "Divide the Pot" game described in the second example because (1) the sum of the payoffs is equal to 30, and (2) the payoff for each player is greater than or equal to zero which is the largest payoff that a player can be sure he will attain if he refuses to join a coalition.

There are usually many imputations for an *n*-person game, even for the simple games such as those used in the examples. In fact, if fractional payoffs are allowed there are an infinite number of imputations for both examples. Thus, our problem is to find a criterion which enables us to determine when one of these imputations is to be preferred to another imputation.

Domination. Assume we are given two different imputations X_1 and X_2. By definition, the sum of the payoffs for any imputation is a constant. Therefore, if the payoff for one player is higher in X_1 than it is in X_2 the payoff for some other player must be lower in X_1 than it is in X_2. Therefore, some players will prefer X_1 to X_2 but other players will prefer X_2 to X_1. That is, it is not possible for all players to prefer X_1 to X_2. However, it is not sufficient to merely state that some players prefer X_1 to X_2. What is necessary is that the players who prefer X_1 are strong enough to enforce the choice of X_1.

Imputation X_1 is said to dominate imputation X_2 for a coalition if the payoffs for all members of the coalition are *greater* for X_1 than for X_2 and if the total payoff for the coalition is large enough to provide the individual payoffs given by X_1. For example, if X_1 = (60, 40, 0) and X_2 = (50, 0, 50), X_1 dominates X_2 for the coalition *AB*. Imputation X_2 does not dominate X_1 for any coalition. If X_3 =

$(0, 50, 50)$, X_3 dominates X_1 for the coalition BC. It should be noted that X_2 does not dominate X_3, nor does X_3 dominate X_2 for any coalition because in each case the payoff for only one player is greater.

Stability. The solution for a game which has only one undominated imputation is said to be stable because the existence of an undominated imputation implies that there is no coalition with both the desire and the power to change the outcome of the game. The existence of more than one undominated imputation does not pose any great difficulties; it simply means that the game has more than one stable outcome. That is, if both X_1 and X_2 are undominated imputations, whenever either of these outcomes is achieved, no coalition will have the desire or power to change the outcome.

Unfortunately, many games have no undominated imputations. In fact, it can be shown that any constant-sum game which has more than one imputation will not have any undominated imputations.[1] Thus it is necessary to seek another solution concept.

Stable Sets. The notion of using a set of imputations as a solution for a game rather than a single imputation was first developed by John von Neumann and Oskar Morgenstern in the *Theory of Games and Economic Behavior*. A set of imputations is called *stable* if it possesses the following properties.

1. No imputation within the set dominates another imputation in the set (internal stability).
2. All imputations outside of the set are dominated by one of the imputations in the set (external stability).

Consider the set of imputations shown in Fig. 6-6 for the "Divide the Pot" game described earlier. None of these imputations dominates

$$(15,15,0)$$
$$(15,0,15)$$
$$(0,15,15)$$

Fig. 6-6. A stable set for "Divide the Pot."

any other imputation in the set because no two players have a reason to prefer any one imputation to another imputation in the set. Thus this set of imputations has internal stability.

By definition, if the payoff vector is an imputation each player must do at least as well as we would have done if he had not joined a coalition. Thus none of the entries in any imputation for this game

[1] For a proof of this theorem, see Guillermo Owen, *Game Theory*, Saunders, Philadelphia, 1968, p. 164.

can be less than zero. Furthermore, the sum of all the entries in an imputation for this game must equal 30. Therefore if two of the entries in an imputation are equal to 15, the third entry must equal zero. The set of imputations given above contains all possible imputations for which two players receive a payoff of 15. For all other imputations, at least two of the entries must be less than 15. Therefore, every other imputation is dominated by one of the three imputations in the set. Thus this set of imputations has external stability.

The concept of stable sets seems to provide a reasonable approach for solving *n*-person games. However, the theory of stable sets does not provide a general proof of the existence or the uniqueness of stable sets. Although at least one *n*-person game has been constructed which has no stable set solutions,[2] the existence of stable sets is usually not a problem. However, uniqueness does present a major difficulty because most *n*-person games possess an immense collection of stable sets.

Let us again consider the "Divide the Pot" game which was described in Example 2. We wish to show that the set S of all imputations in which player A receives a payoff of \$8 and players B and C divide the remaining \$22 in all possible ways is also a stable set for this game. A few of the imputations in S (which has an infinite number of members), are illustrated in Fig. 6-7.

Fig. 6-7. A few of the imputations in the stable set S for the "Divide the Pot" game.

Player A always receives the same payoff so he is indifferent among the imputations in set S. Since the total payoff for players B and C is always the same, both players cannot prefer one specific

[2] W. F. Lucas, "A Game With No Solution," RAND Memorandum RM-5518-PR, RAND Corporation, October 1967.

imputation in S to another specific imputation in S. Thus the set S has internal stability.

If an imputation is not in S, the payoff for player A must either be greater than \$8 or less than \$8. Let us consider imputation X in which the payoff for player A is greater than \$8. Since the sum of the payoffs for players B and C must be less than \$22, players B and C will collectively prefer all of the imputations in S to the imputation X. Furthermore, because S contains all possible splits, there will be at least one imputation in S for which both player B and player C will receive a greater payoff than they receive in imputation X. Thus every imputation for which the payoff for player A is greater than 8 will be dominated by some member of the set S. For example, the imputation (10, 10, 10) is dominated by the imputation (8, 11, 11) which is in S; the imputation (9, 10, 11) is dominated by the imputation $(8, 10^1/_2, 11^1/_2)$ which is also a member of S.

On the other hand, if the payoff for player A is less than 8, he is sure to join either player B or player C to force a return to one of of the imputations in S. For example, the imputation (6, 12, 12) is dominated by (8, 8, 14), (8, 14, 8) and all other members of S for which the payoff to either player B or player C is greater than 12. Therefore the set S also has external stability and is by definition a *stable set*.

It is easy to define two other stable sets which are the same as set S except that the payoff to player B or player C is \$8 and the other two players divide the remaining \$22. Furthermore, a stable set can be defined whenever the fixed payoff to any player is less than $V/2$, where V is the value of the game. Thus if the payoffs can be infinitely divided, there will be an infinite number of stable-set solutions for any constant-sum, n-person game. The existence of such a multiplicity of stable-set solutions presents problems because it is not clear how one stable set is chosen from among all possible sets, nor is it clear how a particular imputation should be chosen from a given stable-set solution.

Von Neumann's Solution. Von Neumann and Morgenstern suggest that a particular stable set reflects a social norm. In its present stage of development, game theory cannot determine which social norm should operate nor can it prescribe how the players in a coalition should split their total payoff. Once society has decided on the standard of behavior, the bargaining ability of the individual players will determine which particular imputation will be chosen.

Despite its weak conclusions, the stable-set solution concept does predict (but does not prove) that negotiations in a simple three-person game will usually result in one of two situations:

1. One player will be offered a fixed amount less than one-half the value of the game and the other two players will split the remaining portion of the payoff in some unspecified way. That is, the solution will be in a stable set such as the set S in Fig. 6-7.
2. Two players will completely exclude the third player and will split the total payoff equally. That is, the solution will be one of the three imputations shown in Fig. 6-6.

Vickrey's Solution. In an unpublished dissertation, Vickrey proposed a modification of the stable set concept which attempts to restrict the sets of imputations which will qualify as solutions. Vickrey's solution is defined as follows: A solution is a stable set with the qualification that any departure from the solution and return to it is necessarily accompanied by damage to one of the players who initiated the departure.[3] The only stable set for the "Divide the Pot" game which satisfies the conditions of Vickrey's solution is the set given in Fig. 6-6 where one of the three players is completely excluded. Although the Vickrey solution is clear-cut for a simple three-person game, the extension of the same idea to an arbitrary, n-person game is difficult.

Many other concepts have been proposed for the solution of n-person games. Among these are the concept of ψ-*stability* developed by R. D. Luce[4] and the *power index* proposed by Shapley.[5] A consideration of these concepts is beyond the scope of this book.

[3] R. D. Luce, and H. Raiffa, *Games and Decisions*, John Wiley, New York, 1957, p. 213.

[4] R. D. Luce, "A Definition of Stability for n-Person Games," *Annals of Mathematics, Vol.* 59 (1954), pp. 357–366.

[5] L. S. Shapley, "A Value for n-Person Games," in H. W. Kuhn and A. W. Tucker. (eds.), *Contributions to the Theory of Games*, II, *Annals of Mathematics Studies*, No. 28, Princeton U. P., 1953.

CHAPTER 7

Nonnegotiable Games

In Chapter 6 we considered games in which the players were able to negotiate with each other and were able to coordinate their actions so as to attain some preferred outcome. In this chapter we will examine games in which the players make their decisions with full knowledge of each other's payoffs but without negotiation.

TWO-PERSON NONNEGOTIABLE GAMES

Consider the game shown in Fig. 7-1. Let us first treat this game as a negotiable game with a status quo point derived from the player's security levels. The maximin strategy for player X is $(\frac{3}{4}, \frac{1}{4})$ and his security level is $\frac{1}{2}$. Strategy Y_2 is a dominating strategy for player Y thus his security level is 2.

The payoff polygon in Fig. 7-1 indicates that we would expect to find the solution on line AB. The equation for this line is

$$Y = \frac{5}{2} - \frac{X}{2} \qquad (7\text{-}1)$$

To use Nash's method to obtain a solution for this game we must find the maximum of the function

$$Z = (X - X_0)\,(Y - Y_0) \qquad (7\text{-}2)$$

By using Eq. 7-1 to substitute for Y, and by substituting the coordinates of the status quo point $(\frac{1}{2}, 2)$ for X_0 and Y_0 we obtain the following function from Eq. 7-2.

$$Z = (X - \tfrac{1}{2})\,(\tfrac{5}{2} - \tfrac{1}{2}X - 2)$$
$$= -\tfrac{1}{2}X^2 + \tfrac{3}{4}X - \tfrac{1}{4} \qquad (7\text{-}3)$$

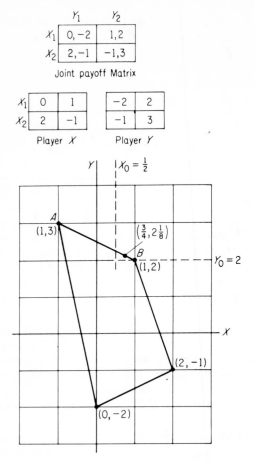

Fig. 7-1.   Nonzero sum game
with one equilibrium point.

We find the value of X which will maximize this function by taking the first derivative and setting it equal to zero:

$$\frac{dZ}{dX} = -X + \tfrac{3}{4} = 0$$

Therefore

$$X^* = \tfrac{3}{4}$$

(From Eq. 7-1),

$$Y^* = \tfrac{5}{2} - \tfrac{1}{2} X^*$$

$$= 2\tfrac{1}{8}$$

Let us now see what would happen if the players made their decisions without negotiation, *but with full knowledge of each other's payoffs*. Since we assume the players are rational, the strategy for player Y is clear; Y_2 dominates Y_1 so player Y will choose strategy Y_2; Player X also knows Y's payoffs so he will conclude that Y will always select Y_2; X's best answer to strategy Y_2 is to play X_1. Thus in the absence of negotiation, the outcome (X_1, Y_2) with payoffs $(1, 2)$ will result.

Because of the decisive role played by the dominating strategy in this game, it is hard to disagree with this solution. The only argument in favor of the solution obtained from the status quo point is that player Y deserves a bonus because his security level is higher than that of player X. However, security levels are determined *in ignorance* of the other player's payoffs by assuming that the players' interests are diametrically opposed. Therefore the security-level solution is difficult to accept if both players are fully aware of all the payoffs for the game they are playing.

THE EQUILIBRIUM POINT

An examination of the outcome (X_1, Y_2) indicates that it has a particular characteristic which makes it a desirable solution. Once the outcome (X_1, Y_2) is obtained, neither player would be willing to move away from this solution by himself if he is making rational decisions. If player X decides to move away he can only move to outcome (X_2, Y_2) which reduces his payoff from 1 to -1. Similarly, if player Y decides to move, the outcome will be (X_1, Y_1) and Y's payoff will decrease from 2 to -2. Neither player is motivated to change his strategy alone for if he does he can only do worse. Such an outcome is called an *equilibrium point*.

The concept of equilibrium is a useful one because it enables us to solve some games without negotiation. J. Nash[1] has shown that every nonzero sum game has at least one equilibrium point. Unfortunately games may have several equilibria. For example, consider the game shown in Fig. 7-2.

The outcomes (X_1, Y_1), (X_2, Y_2) and (X_3, Y_3) are all equilibria since neither player would be motivated to move away from these outcomes by himself. However, the two players have opposite orders of preference with player X preferring (X_1, Y_1) and player Y having a preference for (X_3, Y_3).

Is it reasonable to expect the players to arrive at the compromise

[1] J. F. Nash, "Noncooperative Games," *Annals of Mathematics*, Vol. 54 (1951), pp. 286–295.

	Y_1	Y_2	Y_3
X_1	4,2	0,0	0,0
X_2	0,0	3,3	0,0
X_3	0,0	0,0	2,4

Fig. 7-2. Game with three equilibrium points.

situation, (X_2, Y_2) without negotiation? When players cannot communicate they must depend entirely on guesses about what the other player is going to do. In some cases (such as the game in Fig. 7-1 which has a dominant strategy for player Y) such guesses can be defended on rational grounds. Can a reasonable guess be made about the game in Fig. 7-2?

T. C. Schelling[2] argues that whenever it is of interest to both players to coordinate their choices they will seek a pair of strategies which somehow stand out so that even in the absence of negotiation each player can depend on his competitor's good sense to choose the strategy which offers the opportunity for a *tacit* agreement. Schelling calls such a strategy a *prominent* strategy.

Schelling feels that both players in a game such as in Fig. 7-2 can expect their opponent to select strategy 2 because (1) it is an equilibrium point and (2) among the three equilibrium outcomes in this game, it is the prominent one because of its symmetry. Both of the other strategies favor one of the players but the outcome (X_2, Y_2) is unique because it does not favor either player.

The concept of equilibrium coupled with the notion of prominence seems to extend the theory of nonzero-sum games to the nonnegotiable case. If a game has only one equilibrium point, the corresponding pair of decisions is an acceptable solution by virtue of the characteristics of an equilibrium. If there are several equilibria and one happens to be prominent, then the choices associated with this equilibrium point provide a reasonable solution for a nonnegotiable game. Although this extension of the theory can be defended in many cases there are, unfortunately, many situations in which it leads to paradoxes.

SOLUTION PARADOXES

Figure 7-3 illustrates a two-person nonzero-sum game for which the perfectly rational behavior of game theory can lead to a solution

[2] T. C. Schelling, *The Strategy of Conflict*, Harvard U.P., Cambridge, Mass., 1960.

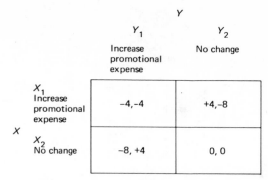

Fig. 7-3. The Sales Manager's Dilemma.

which is intuitively unacceptable. This game is commonly called the "Prisoner's Dilemma," a name derived from the story which is frequently used to illustrate the mixed motives which underlie the game.[3] We will discuss the game in a marketing context.

The sales manager of firm X is considering an increase in the promotional expenditure for a product which competes with only one other product, a very similar product sold by firm Y. He feels that the increased promotional expense will not expand the total market for the product but will increase firm X's share of the market and profit. The profit changes which the sales manager expects to occur are summarized in Fig. 7-3 and are discussed below.

Each firm has two alternatives: to increase the promotional expenditure for the product or to make no change. The three possible outcomes are as follows:

1. If both firms increase their promotional expense there will be no change in the share of the market but both firms will incur a loss in profit equal to the extra promotional expense.
2. If both firms make no change in promotional expense for the product there will be no change in the profit for either firm.
3. If one firm increases its promotional expense, its profit will be greatly increased providing the competing firm does not change its promotional expense. The firm that does not increase promotional expense will suffer a loss in its share of the market and its profit. Thus, the outcomes for strategies (X_1, Y_2) and (X_2, Y_1) are symmetric, with the positive payoff assigned to the firm which increases its promotional expenditure.

The game diagram for the Sales Manager's Dilemma is shown in

[3] An extensive investigation of this game in an experimental context can be found in A. Rapapert and A. Chammah, *Prisoner's Dilemma: A Study of Conflict and Cooperation*, U. of Michigan Press, Ann Arbor, 1965.

	y_1	y_2
x_1	−4, −4	4, −8
x_2	−8, 4	0, 0

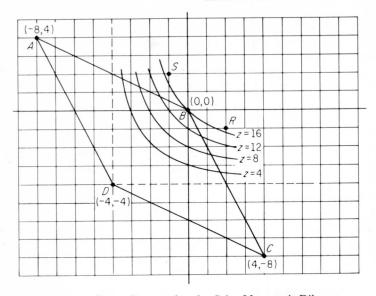

Fig. 7-4. Game diagram for the Sales Manager's Dilemma.

Fig. 7-4. This game illustrates one of the problems which may be encountered in obtaining Nash's solution for a nonzero-sum game. Since strategy 1 is a dominant strategy for both players, the security level for both players is −4. If we use these security levels as the status quo point, the function which has to be maximized to obtain Nash's solution is

$$Z = (X + 4)(Y + 4) \tag{7-4}$$

Since the negotiation set includes portions of both line AB and line BC, two solutions are possible. If line AB is used to obtain the solution.

$$Y = -\tfrac{1}{2}X$$

$$Z = (X + 4)(-\tfrac{1}{2}X + 4)$$

$$= -\tfrac{1}{2}X^2 + 2X + 16$$

$$\frac{dZ}{dX} = -X + 2$$

Therefore the maximum value of Z will occur at the point $(2,-1)$. However, this solution (point R in Fig. 7-4) is outside of the payoff polygon so it is not an acceptable solution for the game.

Similarly, if the line BC is used to obtain Nash's solution, the solution point is $(-1,2)$, point S in Fig. 7-4. This solution is also outside of the payoff polygon so it must also be rejected.

In a situation like this it is necessary to find the solution for the constrained maximum problem; that is we must find the value for X which is inside or on the border of the payoff polygon which gives the largest value for Z. For this particular problem the solution must lie inside or on the boundary of the polygon $ABCD$.

The curved solid lines in Fig. 7-4 are isolines (lines of equal value) for the value of Z. As can be seen in the diagram, the highest value of Z ($Z = 16$) occurs at point B $(0,0)$ which is on the boundary of the payoff polygon. Therefore, Nash's solution for the Sales Manager's Dilemma is the point $(0,0)$ which corresponds to the strategy (X_2,Y_2). It should be noted that this solution is not an equilibrium point.

If the players cannot communicate the game is nonnegotiable and the situation is quite different. Strategy 1 dominates strategy 2 for both players, thus if both players are perfectly rational the outcome will be (X_1,Y_1) and both players will receive a negative payoff. The same result will be obtained from an analysis of the situation facing either of the firms.

Now let us assume that the firms have made a tentative agreement not to change promotional expense. What should the sales manager of firm X do? As indicated by the payoffs in Fig. 7-3, firm X would benefit from breaking the tentative agreement with firm Y regardless of whether Y keeps the agreement or not. If Y keeps the agreement and chooses strategy 2, X can earn the maximum payoff by selecting strategy 1. On the other hand, if firm Y does not keep the agreement there is all the more reason for the sales manager of firm X to do likewise, for if he does not increase the promotional expenditure, his firm will be left holding the bag. Since the payoffs are symmetric, both sides are constantly tempted to break the agreement and the knowledge that the competitor is tempted makes defection practically compelling. If both firms defect the solution is (X_1,Y_1) with payoffs $(-4,-4)$. This solution is an equilibrium point but it is not intuitively satsifactory.

The problems associated with the solution of this game come from the game theorist's concept of a rational player. By definition, a rational player is concerned solely with his own interest. He is not malicious (i.e., he will not select a strategy simply to make his competitor lose) but he does not understand the concept of collective interest. In comparing two courses of action he compares only the expected payoffs which will accrue to him personally. For this reason, a rational player cannot be induced to select strategy 2 in the absence of negotiation. Thus if both players are rational and if negotiations are prohibited both players will select strategy 1 and both will be worse off than if they were guided by their joint interest to select strategy 2. No matter how perfect the logic which leads to it, the solution of the Sales Manager's Dilemma is not intuitively satisfying.

The discrepancy between strategic logic and common sense in the solution of games such as the Prisoner's Dilemma does not exhaust the difficulties encountered with the equilibrium-solution concept. The game in Fig. 7-5 illustrates another type of difficulty.

$$Y$$

	Y_1	Y_2
X_1	(2,2)	(-3,3)
X_2	(3,-3)	(-4,-4)

X

Fig. 7-5. Another game paradox.

This game resembles the Prisoner's Dilemma in that both players are tempted to defect from the cooperative strategy (X_1, Y_1) and if both defect they will both be punished in the outcome (X_2, Y_2). However, this game differs from the Prisoner's Dilemma in that neither player has a dominant strategy. In the absence of negotiation, what strategy should the players of this game adopt?

The game illustrated in Fig. 7-5 has equilibria at both (X_1, Y_2) and (X_2, Y_1). Of these equilibria, player Y prefers the former and player X prefers the latter. If each player wants to maximize his payoff he would choose the strategy which contains his preferred equilibrium point. The resulting outcome, (X_2, Y_2), is not only the worst outcome for both players but is also not intuitively satisfactory.

Each player could also adopt the prudent strategy of making the choice which will guarantee that he will receive his security level. The security level for both players is -3 and is achieved by the selection of strategy 1, thus the solution would be (X_1, Y_1). This solution is intuitively appealing because each player receives a positive payoff and the total payoff is maximized; however, (X_1, Y_1) is not an equilibrium point.

These solutions are derived from different concepts of individual rationality. In the first solution the players attempt to maximize their individual gains; in the latter solution they are trying to minimize their losses. The selection of the "best" solution is clearly dependent on the analyst's definition of rationality.

The concept of collective rationality also leads to two solutions for the game in Fig. 7-5. The first solution, (X_1, Y_1), can be justified by the simple fact that it benefits both players the most. Other arguments which favor this solution will be discussed later.

The second solution which is based on the concept of collective rationality is derived from the use of mixed strategies to determine an equilibrium point for the game. Suppose that player X decides to play a mixed strategy $(p, 1 - p)$ and that player Y adopts a mixed strategy $(q, 1 - q)$. The expected value of the payoff for either player can be computed as shown in Fig. 7-6.

Strategy	Probability of Occurrence	Payoff for X	Expected Value of X's Payoff
X_1, Y_1	pq	2	$2pq$
X_1, Y_2	$p(1-q)$	-3	$3pq - 3p$
X_2, Y_1	$q(1-p)$	3	$-3pq \qquad + 3q$
X_2, Y_2	$(1-p)(1-q)$	-4	$-4pq + 4p \quad + 4q - 4$
	Expected value	=	$-2pq + p + 7q - 4$

Fig. 7-6. The computation of the expected value of the payoff for player X.

The expected value of the payoff for player X can be written as

$$E(X) = p(1 - 2q) + 7q - 4 \qquad (7\text{-}5)$$

A similar computation will show that the expected value of the payoff for player Y is

$$E(Y) = q(1 - 2p) + 7p - 4 \qquad (7\text{-}6)$$

It should be noted that if q is equal to $\frac{1}{2}$, the first term in Eq. 7-5 is equal to zero. Thus if player Y adopts the mixed strategy $(\frac{1}{2}, \frac{1}{2})$ the expected value of the payoff for player X will be $-\frac{1}{2}$ *regardless of the strategy chosen by player X.* Similarly, if player X adopts the mixed strategy $(\frac{1}{2}, \frac{1}{2})$ the first term in Eq. 7-6 will be zero and the expected value of the payoff for player Y will be $-\frac{1}{2}$ regardless of the strategy chosen by Y. Therefore the selection of the strategy $(\frac{1}{2}, \frac{1}{2})$ by *both* players represents an equilibrium point because neither player can improve his own payoff by moving away from this solution point. This solution gives each player a payoff of

$-\frac{1}{2}$ which is above this security level but below the payoff for strategy (X_1, Y_1).

The advocates of the mixed strategy solution point out that the mixed strategy equilibrium point is the only *symmetric* equilibrium point for this game and therefore it is a prominent equilibrium point as defined by Schelling. However, if one chooses symmetry as the basis for selecting a solution, without requiring that the solution be an equilibrium point, the choice is between the strategies (X_1, Y_1), (X_2, Y_2) and the mixed strategy. It is not difficult to eliminate (X_2, Y_2) as a rational strategy but it is very difficult to decide which of the other two symmetric solutions is the "best."

To examine the problems encountered in attempting to select the "best" solution for this game, let us determine what player X can expect to gain from adopting the mixed equilibrium strategy $(\frac{1}{2}, \frac{1}{2})$. If we let p equal $\frac{1}{2}$ in Eq. 7-5 we obtain the following equation for the expected value of the payoff for player X.

$$E(X) = \frac{1}{2}(1 - 2q) + 7q - 4$$
$$= -3\frac{1}{2} + 6q \qquad (7\text{-}7)$$

Equation 7-7 shows that if player X adopts the strategy $(\frac{1}{2}, \frac{1}{2})$ player Y, who controls the value of q, will have complete control of X's payoff. If Y also adopts the mixed strategy $(\frac{1}{2}, \frac{1}{2})$, $E(X) = -\frac{1}{2}$ and the mixed strategy equilibrium solution will result. However, if player Y sets q equal to 0, (i.e., if Y always plays Y_2) player X will receive a payoff of $-3\frac{1}{2}$, which is below his security level, while Y continues to receive a payoff of $-\frac{1}{2}$.

Is it rational for player Y to select a value of q which will depress the payoff for player X? Since Y cannot affect his own payoff, there seems to be no incentive for Y to decrease the payoff for his competitor. Furthermore, X could retaliate by playing X_2; the resulting solution, (X_2, Y_2) is the worst possible outcome for both players. Thus, from this point of view a decision to set q equal to zero would seem to be an irrational action for player Y.

However, an examination of the payoff matrix in Fig. 7-5 indicates that if player X wants to minimize his losses, his best response to setting q equal to zero is to let p equal 1, that is always play X_1. The resulting solution (X_1, Y_2) gives player Y the highest possible payoff he can receive in this game. Thus it can certainly be argued that if Y anticipates the latter response, the decision to set q equal to zero is a rational action.

As indicated by the above discussion, the selection of the best action depends not only on how each player defines rationality but also on how each player perceives his competitor's concept of rationality. For example, if player X believes that prudence is rational he will choose strategy 1. If he believes that player Y has the same concept of rationality he will expect him to be prudent and to choose his first strategy. Thus both players will select strategy 1 and the favorable outcome (X_1, Y_1) will result.

On the other hand, if player X believes it is rational to maximize his gains he will play X_2 whenever Y plays Y_1, and he will expect player Y to play Y_2 whenever he plays X_1. Thus if either player selects his first strategy he would expect his competitor to select the second strategy. If both players believe it is rational to maximize their payoff, neither player will select the first strategy and give his opponent an opportunity to clean up. Therefore, the solution (X_1, Y_1) is not accessible unless negotiations are permitted. However, both players realize that always selecting the second strategy would lead to the worst possible outcome. Thus whenever both players believe that the rational action is to maximize their individual gains, we can expect to obtain the mixed strategy solution.

It should be clear from these examples that neither the maximin nor the equilibrium concept will provide a satisfactory solution for all nonnegotiable games. As we have shown, the choice of the maximin solution for the game illustrated in Fig. 7-3 leads to an equilibrium outcome which is bad for both players. On the other hand, if each player selects the maximin solution for the game in Fig. 7-5 the outcome is satisfactory but it is not an equilibrium point. We must conclude, therefore, that the selection of the "best" solution for a nonnegotiable game depends on the definition of rationality and that a single definition of rationality may not be sufficient for the analysis of all nonnegotiable, nonzero-sum games.

SUCCESSIVE PLAYS OF A GAME

It is reasonable to think that the analysis of a nonnegotiable game might be different if the players knew that the game was to be played many times in succession. For example, in the Sales Manager's Dilemma a player might hesitate to play his first strategy if he knew that the game was to be played many times because this action might induce his competitor to retaliate on a successive play of the game. On the other hand, it seems sensible for a player to select strategy 2 as a way of communicating to his competitor that he is willing to cooperate as long as his opponent will continue to select the second strategy. Each player is prevented from selecting strategy 1 by the knowledge that his competitor may retaliate on a successive play by also playing strategy 1. Thus the players will adopt the solution

(X_2, Y_2) by *tacit* agreement. It seems reasonable for the players to stick with this solution until the very last play of the game after which no retaliation is possible.

Unfortunately, the exemption of the last play from the tacit agreement proves to be the downfall of the entire argument. For example, if the game is to be played 25 times, on the last play it would be advantageous for player X, for example, to select strategy 1 regardless of which strategy Y selects on the last move. Since similar reasoning applies to player Y, both players know that on the last play of the game the outcome will be (X_1, Y_1). But if the last outcome is known unconditionally, the twenty-fourth play becomes the last effective play of the game. The same reasoning that was used to analyze the twenty-fifth play can be applied to the twenty-fourth play so that the twenty-fourth outcome will also be an unconditional (X_1, Y_1). Thus the entire system of tacit collusion, which appears to make strategic sense, collapses and we again arrive at the conclusion that players who are pursuing their individual interests ought to select their first strategy all 25 times. This solution will result in a loss of 100 for each player whereas playing the second strategy all 25 times in succession would have resulted in no loss for either player.

Even though the choice of strategy 1 in a single play of the game may seem to be justified, the choice of this strategy 25 times in succession is difficult to accept when an opportunity exists to establish a tacit agreement between the players.

NONNEGOTIABLE *n*-PERSON GAMES

There is no great difference between the theory of nonnegotiable *n*-person games and nonnegotiable two-person games. It can be shown that any finite nonnegotiable *n*-person game has at least one equilibrium *n*-tuple of mixed strategies. All the conceptual difficulties which are encountered in determining equilibrium points for two-person games are also present in *n*-person games. However, the computation of mixed strategy equilibrium *n*-tuples is much more difficult than the computation of equilibrium pairs. In addition, the existence of more than two players does nothing to alleviate the problems inherent in the definition of rationality. Thus *n*-person games are more complex computationally and are potentially more complex if multiple definitions of rationality are permitted.

SUMMARY

In view of the situation described in this chapter, can we speak of a general theory for nonnegotiable games? It should be clear from the examples discussed above that the analysis of some nonnegotiable games will lead to paradoxical results if the zero-sum game definition of rationality is used. To avoid this problem we must use more than one definition of rationality or we must accept results which are intuitively unsatisfactory. The lack of a single suitable definition of rationality does not hurt the purely formal aspects of game theory. In fact, the recognition of the ambivalent nature of rationality in the analysis of nonnegotiable games is a valuable discovery. However, the prescriptive aspects of game theory are hurt by the problems encountered in defining rationality because the theory cannot prescribe a rational decision if the meaning of rationality is not clear.

Bibliography

BOOKS

Bennion, Edward G., *Elementary Mathematics of Linear Programming and Game Theory*. East Lansing: Bureau of Business and Economic Research, College of Business and Public Service, Michigan State University, 1960.

Berge, C., and A. Ghouila-Houri, *Programming, Games and Transportation Networks*. New York: John Wiley & Sons, 1965.

Blackwell, D., "Game Theory," in J. F. McCloskey and F. N. Trefethen (eds.), *Operation Research for Management*. Baltimore: Johns Hopkins Press, 1954.

————, and M. A. Girschick, *Theory of Games and Statistical Decisions*. New York: John Wiley & Sons, 1954.

Bohnenblust, H. F., "The Theory of Games," in E. F. Beckenback (ed.), *Modern Mathematics for the Engineer*. New York: McGraw-Hill Book Company, 1956.

Boore, William F., "Game Theory and the Behavior of the Firm," in Joseph W. McGuire, (ed.), *Interdisciplinary Studies in Business Behavior*. Cincinnati: Southwestern Publishing Company, 1962.

Braithwaite, R. B., *Theory of Games as a Tool for the Moral Philosopher*. Cambridge: Cambridge University Press, 1955.

Burger, Ewald, *Introduction to the Theory of Games*. Englewood Cliffs, N.J.: Prentice-Hall, 1963.

Dresher, Melvin, *Games of Strategy: Theory and Applications*. Englewood Cliffs, N.J.: Prentice-Hall, 1961.

————, L. S. Shapley, and A. W. Tucker (eds.), *Advances in Game Theory*, *Annals of Mathematics Studies No. 52*. Princeton: University Press, 1964.

————, A. W. Tucker, and P. Wolfe, (eds.), *Contributions to the Theory of Games*, *Vol. III*, *Annals of Mathematics Studies No. 39*. Princeton University Press, 1957.

Gale, David, *The Theory of Matrix Games and Linear Economic Models.* Providence: Department of Mathematics, Brown University, 1957.

Glicksman, Abraham M., *An Introduction to Linear Programming and the Theory of Games.* New York: John Wiley & Sons, 1963.

Isaacs, Rufus Philip, *Differential Games: A Mathematical Theory with Applications to Warfare and Pursuit,* Control and Optimization. New York: John Wiley & Sons, 1965.

Karlin, Samuel, *Mathematical Methods and Theory in Games, Programming, and Economics,* Vol. I and II. Reading, Mass.: Addison-Wesley Publishing Company, 1959.

Kuhn, H. W., *Lectures on the Theory of Games.* Princeton, N.J.: Princeton University Press, 1952.

———, and A. W. Tucker (eds.), *Linear Inequalities and Related Systems, Annals of Mathematics Studies No. 38.* Princeton: Princeton University Press, 1956.

———, *Contributions to the Theory of Games,* Vols. I and II, *Annals of Mathematics Studies Nos. 24, 28.* Princeton: Princeton University Press, 1950, 1953.

Luce, R. D., and H. Raiffa, *Games and Decisions.* New York: John Wiley & Sons, 1957.

McDonald, J., *Strategy in Poker, Business and War.* New York: W. W. Norton & Company, 1950.

McKinsey, John Charles Chenoweth, *Introduction to the Theory of Games.* New York: McGraw-Hill Book Company, 1952.

Owen, Guillermo, *Game Theory.* Philadelphia: W. B. Saunders Company, 1968.

Rapoport, Anatol, *Fights, Games, and Debates.* Ann Arbor: University of Michigan Press, 1960.

———, *Strategy and Conscience.* New York: Harper & Row, Publishers, 1964.

———, *Two-Person Game Theory: The Essential Ideas.* Ann Arbor: University of Michigan Press, 1966.

———, and Albert M. Chammah, *Prisoner's Dilemma: A Study of Conflict and Cooperation.* Ann Arbor: University of Michigan Press, 1965.

Schelling, T. C., *The Strategy of Conflict.* Cambridge, Mass.: Harvard University Press, 1960.

Shubik, Martin (ed.), *Readings in Game Theory and Political Behavior.* Garden City, N.Y.: Doubleday & Company, 1954.

———, *Game Theory and Related Approaches to Social Behavior.* New York: John Wiley & Sons, 1964.

Snyder, R., "Game Theory and the Analysis of Political Behavior," in Brookings Lectures, *Research Frontiers in Politics and Government*. Washington, D.C.: Brookings Institution, 1955, pp. 70–103.

Thompson, Gerald Luther, *Lectures on Game Theory, Markov Chains and Related Topics*. Work performed under AEC Contract AT(29-1)789. Albuquerque, N.M.: Sandia Corporation, 1958.

Tucker, A. W., and R. D. Luce (eds.), *Contributions to the Theory of Games*, Vol. IV. *Annals of Mathematics Studies No. 40*. Princeton: Princeton University Press, 1959.

Vajda, S., *The Theory of Games and Linear Programming*. New York: John Wiley & Sons, 1956.

————, *An Introduction to Linear Programming and the Theory of Games*. New York: John Wiley & Sons, 1960.

Venttsel, E. S., *Elements of the Theory of Games*. New York: Gordon and Breach, Science Publishers, Inc., 1959.

————, *Lectures on Game Theory*. New York: Gordon and Breach, Science Publishers, 1961.

————, *An Introduction to the Theory of Games*. Boston: D.C. Heath and Company, 1963.

Von Neumann, J., and O. Morgenstern, *Theory of Games and Economic Behavior*. 3d ed. New York: John Wiley & Sons, 1964.

Williams, J. D., *The Compleat Strategyst*. 2d ed. New York: McGraw-Hill Book Company, 1966.

PERIODICALS

Arrow, K. J., M. J. Beckman, and S. Karlin, "Game Theory Methods Applied to the Optimum Expansion of the Capacity of a Firm," *Stanford University Department of Economics Technical Report 27*, November 1955.

Aumann, R. J., "Almost Strictly Competitive Games," *S.I.A.M. Journal*, Vol. 9 (1961), pp. 544–550.

Baligh, H. H., and L. E. Richartz, "Variable-Sum Game Models of Marketing Problems," *Journal of Marketing Research*, Vol. 4 (1967), pp. 173–183.

Bellman, R., "The Theory of Games," *U.S. Government Research Reports AD-606 365*. Washington, D.C.: Clearinghouse, Department of Commerce, 1957.

Berkovitz, L. D., and M. Dresher, "Game-Theory Analysis of Tactical Air War," *Operations Research*, Vol. 7 (1959), pp. 599–620.

————, "Allocation of Two Types of Aircraft in Tactical Air War, A Game Theoretic Analysis," *Operations Research*, Vol. 8 (1960), pp. 694–706.

Berkman, H. G., "Game Theory of Land Use Determination," *Land Economics*, Vol. 41 (1965), pp. 11–19.

Blackwell, D., "Game Theory for War Gaming," *U.S. Government Research Reports PB-165 227 and AD-236 152*. Washington, D.C.: Clearinghouse, Department of Commerce, 1957.

Brown, R. H., "The Solution of a Certain Two-Person Zero-Sum Game," *Operations Research*, Vol. 5 (1957), p. 63.

Charnes, A., M. Kirby, and W. Raike, "Chance-Constrained Games with Partially Controllable Strategies," *Operations Research*, Vol. 16 (1968), pp. 142–149.

Dalkey, N. C., "Games and Simulations," *U.S. Government Research Reports AD-601 138*. Washington, D.C.: Clearinghouse, Department of Commerce, 1964.

Danskin, J. M., "A Game Theory Model of Convoy Routing," *Operations Research*, Vol. 10 (1962), pp. 774–784.

Davis, M., and M. Maschler, "The Kernel of a Cooperative Game," *U.S. Government Research Reports AD-418 434*. Washington, D.C.: Clearinghouse, Department of Commerce, 1963.

Dickinson, R., "Game Theory and the Department Store Buyer," *Retailing*, Vol. 42 (1966), pp. 14–24.

Dresher, M., "Methods of Solution in Game Theory," *U.S. Government Research Reports AD-603 837*. Washington, D.C.: Clearinghouse, Department of Commerce, 1949.

———, "Statistical Games," *Mathematics Magazine* (1951), pp. 93–99.

———, "Theory of Games of Strategy," *U.S. Government Research Reports AD-605 125*. Washington, D.C.: Clearinghouse, Department of Commerce, 1956.

Firstman, S. I., "A Game Theoretic Approach to Space Vehicle Prelaunch Activities Prescheduling," *U.S. Government Research Reports AD-604 454*. Washington, D.C.: Clearinghouse, Department of Commerce, 1962.

Friedman, L., "Game Theory Models in the Allocation of Advertising Expenditures," *Operations Research*, Vol. 6 (1958), pp. 699–708.

Gillman, L., "Operations Analysis and the Theory of Games: An Advertising Example," *Journal of the American Statistical Association* (1950).

Glasser, G. J., "Personnel Decisions and the Theory of Games," *Personnel Administration*, Vol. 21 (1958), pp. 6–11.

Gluss, B., "A Bayesian Solution to a Sequential Two-Person Zero-Sum Game," *U.S. Government Research Reports PB 166-106*. Washington, D.C.: Clearinghouse, Department of Commerce, 1963.

Hale, I., "Theory of Games in Stock Selection, How It Might Be Applied to Security Analysis," *Financial Analysts Journal*, Vol. 16 (1960) pp. 53–56.

Harsanyi, J. C., "On the Rationality Postulates Underlying the Theory of Cooperative Games," *Journal of Conflict Resolution*, Vol. 5 (1961), pp. 179–196.

————, "Measurement of Social Power, Opportunity Costs, and the Theory of Two-Person Bargaining Games," *Behavioral Science*, Vol. 7 (1962), pp. 67–80.

————, "Measurement of Social Power in n-Person Reciprocal Situations," *Behavioral Science*, Vol. 7 (1962), pp. 81–91.

————, "A General Theory of Rational Behavior in Game Situations," *Econometrica* Vol. 34 (1966), pp. 613–634.

————, "Games with Incomplete Information Played by 'Bayesian' Players." Part I; The Basic Model; Part II, Bayesian Equilibrium Points; Part III, the Basic Probability Distribution of the Game. *Management Science*, Vol. 14 (1967–68), pp. 159–182, 320–334, 486–502.

Haywood, O. G., Jr., "Military Decision and Game Theory," *Operations Research*, Vol. 2 (1954), pp. 365–385.

Heiss, K., "Game Theory and Human Conflict," *Economic Research Program Research Memorandum No. 80*. Princeton: Princeton University Press, 1966.

Hoffman, A. J., and R. M. Karp, "On Nonterminating Stochastic Games," *Management Science*, Vol. 12 (1966), pp. 359–370.

Hurwicz, L., "Game Theory and Decisions," *Scientific American* (1955), pp. 78–83.

Jentzch, G., "Some Thoughts on the Theory of Cooperative Games," *U.S. Government Research Reports AD-401 642*. Washington, D.C.: Clearinghouse, Department of Commerce, 1963.

Kahn, H., and I. Mann, "Game Theory," *U.S. Government Research Reports AD-606 530*. Washington, D.C.: Clearinghouse, Department of Commerce, 1957.

Kao, R. C., and J. Staudhammer, "Techniques of Systems Analysis and Design: Game Theory," *U.S. Government Research Reports AD-288 823*. Washington, D.C.: Clearinghouse, Department of Commerce, 1962.

Koo, A. Y. C., "Recurrent Objections to the Minimax Strategy," *Review of Economics and Statistics*, Vol. 41 (1959), pp. 36–41.

Kulik, B. J., "A Comparison of Theoretical Behavior with Observed Behavior in a Two-Person Zero-Sum Game Interaction," *U.S. Government Research Reports AD-618 117*. Washington, D.C.: Clearinghouse, Department of Commerce, 1965.

Lemke, C. E., and J. T. Howson, Jr., "Equilibrium Points of

Bimatrix Games," *S.I.A.M. Journal*, Vol. 12 (1964), pp. 413–423.

McDonald, John, "The Theory of Strategy," *Fortune*, Vol. 34 (1949), pp. 100–110.

Mangasarian, O. L., and H. Stone, "Two-Person Nonzero-Sum Games and Quadratic Programming," *Journal of Mathematical Analysis and Applications*, Vol. 9 (1964), pp. 348–355.

Maschler, M., "The Power of a Coalition," *U.S. Government Research Reports Ad-297 369*. Washington, D.C.: Clearinghouse, Department of Commerce, 1963.

———, "Playing an *n*-Person Game, an Experiment," *U.S. Government Research Reports AD-613 034*. Washington, D.C.: Clearinghouse, Department of Commerce, 1965.

Mayberry, J. P., J. F. Nash, and M. Shubik, "A Comparison of Treatments of a Duopoly Situation," *Econometrica*, Vol. 21 (1953), pp. 141–154.

Messick, D. M., "Bayesian Decision Theory, Game Theory and Group Problem Solving," *U.S. Government Research Reports AD-614 683*. Washington, D.C.: Clearinghouse, Department of Commerce, 1964.

Miyasawa, D., "The *n*-Person Bargaining Game," *U.S. Government Research Reports AD-256 522*. Washington, D.C.: Clearinghouse, Department of Commerce, 1961.

Moglewer, S., "A Game Theory Model for Agricultural Crop Selection," *Econometrica*, Vol. 30 (1962), pp. 253–266.

Morgenstern, O., "The Theory of Games," *Scientific American* (1949), pp. 22–25.

———, "Game Theory in U.S.," excerpt from "Question of National Defense, *Fortune*, Vol. 60 (1959), p. 126.

Nash, J. F., "The Bargaining Problem," *Econometrica*, Vol. 18 (1950), pp. 155–162.

———, "Equilibrium Points in *n*-Person Games," *Proc. Nat. Acad. Sci.*, Vol. 36 (1950), pp. 48–49.

———, "Noncooperative Games," *Annals of Mathematics*, Vol. 54 (1951), pp. 268–298.

———, "Two-Person Cooperative Games," *Econometrica*, Vol. 21 (1953), pp. 128–140.

Nelson, W. G., "Could Game Theory Aid Capital Budgeting?," *N.A.A. Bulletin*, Vol. 43 (1962), pp. 49–58.

Nering, E. D., "Coalition Bargaining in *n*-Person Games." *U.S. Government Research Reports AD-267 309*. Washington, D.C.: Clearinghouse, Department of Commerce, 1961.

Neuts, M., "Games on the Unit-Square with Discrete Payoff," *U.S. Government Research Reports AD-253 011, PB-150 416.* Washington, D.C.: Clearinghouse, Department of Commerce, 1961.

Newman, D. J., "Model for Real Poker," *Operations Research*, Vol. 7 (1959), pp. 557–560.

Owen, G., "A Elementary Proof of the Minimax Theorem," *Management Science*, Vol. 13 (1967), p. 765.

Peleg, B., "Solutions to Cooperative Games Without Side Payments," *Trans. Amer. Math. Soc.*, Vol. 106 (1963), pp. 280–292.

Rapoport, A., "Use and Misuse of Game Theory," *Scientific American*, Vol. 207 (1962), pp. 108–114.

———, "Escape from Paradox," *Scientific American*, Vol. 217 (1967), pp. 50–56.

———, and Carol Orwant, "Experimental Games: A Review," *Behavioral Science*, Vol. 7 (1962), pp. 1–37.

———, A. Chammak, J. Dwyer, and J. Gyr, "Three-Person, Non-zero-Sum, Nonnegotiable Games," *Behavioral Science*, Vol. 7 (1962), pp. 38–58.

Raun, D. L., "Profit Planning and Game Theory," *Management Accounting*, Vol. 47 (1966), pp. 3–10.

Reichardt, R., "Competition Through the Introduction of New Products," *U.S. Government Research Reports AD-261 591.* Washington, D.C.: Clearinghouse, Department of Commerce, 1961.

Robinson, F. D., "Advertising Budget: Using the Game Theory to Determine How Much to Spend," *Controller*, Vol. 26 (1958), p. 368.

Robinson, Julia, "Iterative Method of Solving a Game," *Annals of Mathematics*, Vol. 54 (1951), pp. 296–301.

Rosen, J. B., "Existence and Uniqueness of Equilibrium Points for Concave *n*-Person Games," *Econometrica*, Vol. 33 (1965), pp. 520–534.

Scharf, H. E., "An Analysis of Markets with a Large Number of Participants," *U.S. Government Research Reports AD-268 407*, TISTP/MFA. Washington, D.C.: Clearinghouse, Department of Commerce, 1961.

———, "The Cone of an *n*-Person Game," *Econometrica*, Vol. 35 (1967), pp. 50–69.

Schelling, T. C., "The Strategy of Conflict: Prospectus for a Reorientation of Game Theory." *Conflict Resolution*, Vol. 2 (1958), pp. 203–264.

————, "For the Abandonment of Symmetry in Game Theory," *Review of Economics and Statistics*, Vol. 41 (1959), pp. 213–224.

Scopel, A., "Induced Collaboration in Some Nonzero-Sum Games," *Journal of Conflict Resolution*, Vol. 6 (1962), pp. 335–340.

Shakun, M. F., "Advertising Expenditures in Coupled Markets—A Game Theory Approach," *Management Science*, Vol. 11 (1966), pp. B42–B47.

————, Dynamic Model for Competitive Marketing in Coupled Markets (Use of Game-Theory Methods)," *Management Science*, Vol. 12 (1966), pp. B525–B530.

Shapley, L. S., "Stochastic Games," *Proc. Nat. Acad. Sci.*, Vol. 39 (1953), pp. 327–332.

————, "Values of Large Games III: A Corporation with Two Large Stockholders," *U. S. Government. Research Reports AD-269 105*, Washington, D.C.: Clearinghouse, Department of Commerce, 1961.

————, "Simple Games: An Outline of the Descriptive Theory," *Behavioral Science*, Vol. 7 (1962), pp. 59–66.

————, "Some Topics in Two-Person Games," *U. S. Government Research Reports, AD-407, 145*. Washington, D.C.: Clearinghouse, Department of Commerce, 1962.

————, "Notes on *n*-Person Games VII, Cores of Convex Games," *U. S. Government Research Reports AD-624 310*. Washington, D.C.: Clearinghouse, Department of Commerce, 1965.

Shubik, M., "Information, Theories of Competition, and the Theory of Games," *Journal of Political Economy* (1952), pp. 145-150.

————, "The Uses of Game Theory in Management Science," *Management Science*, Vol. 1 (1955), pp. 32ff.

————, "Some Reflections on the Design of Game Theoretic Models for the Study of Negotiation and Threats," *Journal of Conflict Resolution*, Vol. 7 (1963), pp. 1-12.

Sisson, R. L., "Games (Use in Operations Research and Management Science)," *Systems and Procedures Magazine*, Vol. 12 (1961), pp. 32-36.

Smith, S. B., "Game Theory: New Tool for P.A.'s," *Purchasing*, Vol. 50 (1961), pp. 86-89.

Thompson, G., "Game Theory and 'Social Value' States," *Ethics*, Vol. 75 (1964), pp. 36–39.

Tukey, J. W., "A Problem of Strategy," *Econometrica*, Vol. 17 (1949).

Wald, Abraham, "Review of the Theory of Games," *Review of Economics and Statistics*, Vol. 34 (1947), pp. 47-52.

Wang, H., "Games, Logic, and Computers," *Scientific American*, Vol. 213 (1965), pp. 98–104.

Weiss, H. K., "Some Differential Games of Tactical Interest and the Value of a Supporting Weapon System," *Operations Research*, Vol. 7, (1959), pp. 180–196.

Wilson, K. V., and V. E. Bixenstine, "Formes of Social Control in Two-Person, Two-Choice Games," *Behavioral Science*, Vol. 7 (1962), pp. 92–102.

Index